LONDON MATHEMATICAL SOCIETY LECTURE NOTE SERIES

1. General cohomology theory and K-theory, P.HILTON
4. Algebraic topology, J.F.ADAMS
5. Commutative algebra, J.T.KNIGHT
8. Integration and harmonic analysis on compact groups, R.E.EDWARDS
9. Elliptic functions and elliptic curves, P.DU VAL
10. Numerical ranges II, F.F.BONSALL & J.DUNCAN
11. New developments in topology, G.SEGAL (ed.)
12. Symposium on complex analysis, Canterbury, 1973, J.CLUNIE & W.K.HAYMAN (eds.)
13. Combinatorics: Proceedings of the British Combinatorial Conference 1973, T.P.McDONOUGH & V.C.MAVRON (eds.)
15. An introduction to topological groups, P.J.HIGGINS
16. Topics in finite groups, T.M.GAGEN
17. Differential germs and catastrophes, Th.BROCKER & L.LANDER
18. A geometric approach to homology theory, S.BUONCRISTIANO, C.P. ROURKE & B.J.SANDERSON
20. Sheaf theory, B.R.TENNISON
21. Automatic continuity of linear operators, A.M.SINCLAIR
23. Parallelisms of complete designs, P.J.CAMERON
24. The topology of Stiefel manifolds, I.M.JAMES
25. Lie groups and compact groups, J.F.PRICE
26. Transformation groups: Proceedings of the conference in the University of Newcastle-upon-Tyne, August 1976, C.KOSNIOWSKI
27. Skew field constructions, P.M.COHN
28. Brownian motion, Hardy spaces and bounded mean oscillations, K.E.PETERSEN
29. Pontryagin duality and the structure of locally compact Abelian groups, S.A.MORRIS
30. Interaction models, N.L.BIGGS
31. Continuous crossed products and type III von Neumann algebras, A.VAN DAELE
32. Uniform algebras and Jensen measures, T.W.GAMELIN
33. Permutation groups and combinatorial structures, N.L.BIGGS & A.T.WHITE
34. Representation theory of Lie groups, M.F. ATIYAH et al.
35. Trace ideals and their applications, B.SIMON
36. Homological group theory, C.T.C.WALL (ed.)
37. Partially ordered rings and semi-algebraic geometry, G.W.BRUMFIEL
38. Surveys in combinatorics, B.BOLLOBAS (ed.)
39. Affine sets and affine groups, D.G.NORTHCOTT
40. Introduction to Hp spaces, P.J.KOOSIS
41. Theory and applications of Hopf bifurcation, B.D.HASSARD, N.D.KAZARINOFF & Y-H.WAN
42. Topics in the theory of group presentations, D.L.JOHNSON
43. Graphs, codes and designs, P.J.CAMERON & J.H.VAN LINT
44. Z/2-homotopy theory, M.C.CRABB
45. Recursion theory: its generalisations and applications, F.R.DRAKE & S.S.WAINER (eds.)
46. p-adic analysis: a short course on recent work, N.KOBLITZ
47. Coding the Universe, A.BELLER, R.JENSEN & P.WELCH
48. Low-dimensional topology, R.BROWN & T.L.THICKSTUN (eds.)

49. Finite geometries and designs, P.CAMERON, J.W.P.HIRSCHFELD & D.R.HUGHES (eds.)
50. Commutator calculus and groups of homotopy classes, H.J.BAUES
51. Synthetic differential geometry, A.KOCK
52. Combinatorics, H.N.V.TEMPERLEY (ed.)
53. Singularity theory, V.I.ARNOLD
54. Markov processes and related problems of analysis, E.B.DYNKIN
55. Ordered permutation groups, A.M.W.GLASS
56. Journées arithmétiques 1980, J.V.ARMITAGE (ed.)
57. Techniques of geometric topology, R.A.FENN
58. Singularities of smooth functions and maps, J.MARTINET
59. Applicable differential geometry, M.CRAMPIN & F.A.E.PIRANI
60. Integrable systems, S.P.NOVIKOV et al.
61. The core model, A.DODD
62. Economics for mathematicians, J.W.S.CASSELS
63. Continuous semigroups in Banach algebras, A.M.SINCLAIR
64. Basic concepts of enriched category theory, G.M.KELLY
65. Several complex variables and complex manifolds I, M.J.FIELD
66. Several complex variables and complex manifolds II, M.J.FIELD
67. Classification problems in ergodic theory, W.PARRY & S.TUNCEL
68. Complex algebraic surfaces, A.BEAUVILLE
69. Representation theory, I.M.GELFAND et al.
70. Stochastic differential equations on manifolds, K.D.ELWORTHY
71. Groups - St Andrews 1981, C.M.CAMPBELL & E.F.ROBERTSON (eds.)
72. Commutative algebra: Durham 1981, R.Y.SHARP (ed.)
73. Riemann surfaces: a view towards several complex variables, A.T.HUCKLEBERRY
74. Symmetric designs: an algebraic approach, E.S.LANDER
75. New geometric splittings of classical knots (algebraic knots), L.SIEBENMANN & F.BONAHON
76. Linear differential operators, H.O.CORDES
77. Isolated singular points on complete intersections, E.J.N.LOOIJENGA
78. A primer on Riemann surfaces, A.F.BEARDON
79. Probability, statistics and analysis, J.F.C.KINGMAN & G.E.H.REUTER (eds.)
80. Introduction to the representation theory of compact and locally compact groups, A.ROBERT
81. Skew fields, P.K.DRAXL
82. Surveys in combinatorics: Invited papers for the ninth British Combinatorial Conference 1983, E.K.LLOYD (ed.)
83. Homogeneous structures on Riemannian manifolds, F.TRICERRI & L.VANHECKE
84. Finite group algebras and their modules, P.LANDROCK
85. Solitons, P.G.DRAZIN
86. Topological topics, I.M.JAMES (ed.)
87. Surveys in set theory, A.R.D.MATHIAS (ed.)

London Mathematical Society Lecture Note Series: 86

Topological Topics

Articles on algebra and topology presented to Professor P.J. Hilton in celebration of his sixtieth birthday

I.M. JAMES
Mathematical Institute, University of Oxford

CAMBRIDGE UNIVERSITY PRESS
Cambridge
London New York New Rochelle
Melbourne Sydney

Published by the Press Syndicate of the University of Cambridge
The Pitt Building, Trumpington Street, Cambridge CB2 1RP
32 East 57th Street, New York, NY 10022, USA
296 Beaconsfield Parade, Middle Park, Melbourne 3206, Australia

First published 1983

Library of Congress catalogue card number: 83-7745

British Library Cataloguing in Publication Data

Topological topics - (Lecture note series/London
Mathematical Society, ISSN 0076-0552; 86)
1. Topology - Addresses, essays, lectures
I. James, I.M. II. Hilton, P.J.
III. Series
514 QA611.15

ISBN 0 521 27581 4

Transferred to digital printing 2002

In honour of Peter Hilton
on the occasion of
his sixtieth birthday

CONTENTS

PUBLICATIONS OF P.J. HILTON

* Book

[1] Calculating the homotopy groups of A_n^2-polyhedra I, Quart. J. Math. Oxford (2), 1 (1950), 299-309.

[2] Calculating the homotopy groups of A_n^2-polyhedra II, Quart. J. Math. Oxford (2), 2 (1951), 228-240.

[3] Suspension theorems and the generalized Hopf invariant, Proc. London Math. Soc. (3), 1 (1951), 462-493.

[4] The Hopf invariant and homotopy groups of spheres, Proc. Cambridge Phil. Soc., 48 (1952), 547-554.

[5*] An introduction to homotopy theory, Cambridge University Press, (1953).

[6] (with M.G. Barratt), On join operations in homotopy groups, Proc. London Math. Soc. (3), 1 (1953), 430-445.

[7] (with J.H.C. Whitehead), Note on the Whitehead product, Ann. of Math., 58 (1953), 429-442.

[8] On the Hopf invariant of a composition element, J. London Math. Soc. 29 (1954), 165-171.

[9] A certain triple Whitehead product, Proc. Cambridge Phil. Soc., 50 (1954), 189-197.

[10] On the homotopy groups of the union of spheres, Proc. Int. Cong. Amsterdam (1954).

[11] On the homotopy groups of the union of spaces, Comment. Math. Helv. 29 (1955), 59-92.

[12] Note on the P-homomorphism in homotopy groups of spheres, Proc. Cambridge Phil. Soc. 51 (1955), 230-233.

[13] On the homotopy groups of the union of spheres, J. London Math. Soc., 30 (1955), 154-172.

[14] Remark on the factorization of spaces, Bull. Acad. Polon. Sci. (1) III. 3 (1955) 579-581.

[15] On Borsuk dependence and duality, Bull. Soc. Math. Belg., 7 (1955), 143-155.

[16] (with J.F. Adams), On the chain algebra of a loop space, Comment. Math. Helv. 30 (1956), 305-330.

[17] Note on the higher Hopf invariants, Proc. Cambridge Phil. Soc. 52 (1956), 750-752.

[18] Remark on the tensor product of modules, Bull.Acad. Polon. Sci. Cl. III., 4 (1956), 325-328.

[19] Generalizations of the Hopf invariant, Colloque de Topologie Algébrique, C.B.R.M. (1956), 9-27.

[20] Note on quasi Lie rings, Fund. Math. 43 (1956), 203-237.

[21] On divisors and multiples of continuous maps, Fund. Math. 43 (1957), 358-386.

[22] (with W. Ledermann) Homology and ringoids I, Proc. Cambridge Phil. Soc., 54 (1958), 152-167.

[23] (with B. Eckmann) Groupes d'homotopie et dualité. Groupes absolus, C.R. Acad. Sci. Paris, 246 (1958), 2444-2447.

[24] (with B. Eckmann) Groupes d'homotopie et dualité. Suites exactes, C.R. Acad. Sci. Paris, 246 (1958), 2555-2558.

[25] (with B. Eckmann) Groupes d'homotopie et dualité. Coefficients, C.R. Acad. Sci. Paris, 246 (1958), 2991-2993.

[26] (with B. Eckmann) Transgression homotopique et cohomologique, C.R. Acad. Sci. Paris, 247 (1958), 629-632.

[27] (with W. Ledermann) Homological ringoids, Coll.Math. 6 (1958), 177-186.

[28*] Differential calculus, Routledge and Kegan Paul, (1958).

[29] (with B. Eckmann) Décomposition homologique d'un polyèdre simplement connexe, C.R. Acad. Sci. Paris, 248 (1959), 2054-2056.

[30] (with B. Eckmann) Homology and homotopy decomposition of continuous maps, Proc. Nat. Acad. Sci., 45 (1959), 372-375.

[31] Homotopy theory of modules and duality, Proc. Mexico Symposium, (1958), 273-281.

[32] (with W. Ledermann) Homology and ringoids II, Proc. Cambridge Phil. Soc. 55 (1959), 129-164.

[33] (with T. Ganea) Decomposition of spaces in cartesian products and unions, Proc. Cambridge Phil. Soc., 55 (1959), 248-256.

[34] (with W. Ledermann) Homology and ringoids III, Proc. Cambridge Phil. Soc., 56 (1960), 1-12.

[35] (with W. Ledermann) On the Jordan-Hölder theorem in homological monoids, Proc. London Math. Soc. (3), 10 (1960), 321-334.

[36] On an generalization of nilpotency to semi-simplicial complexes, Proc. London Math. Soc. (3), 10 (1960), 604-622.

[37] (with B. Eckmann) Operators and co-operators in homotopy theory, Math. Ann. 141 (1960), 1-21.

[38] (with I. Berstein) Category and generalized Hopf invariants, Ill. J. Math., 4 (1960), 437-451.

[39*] Partial derivatives, Routledge and Kegan Paul, (1960).

[40*] (with S. Wylie) Homology Theory, Cambridge Univ. Press, (1960).

[41] (with E.H. Spanier) On the embeddability of certain complexes in Euclidean spaces, Proc. Amer. Math. Soc., 11 (1960), 523-526.

[42] Remark on the free product of groups, Trans. Amer. Math. Soc., 96 (1960), 478-488.

[43] (with W. Ledermann) Remark on the l.c.m. in homological ringoids, Quart. J. Math. Oxford, 11 (1960), 287-294.

[44] (with B. Eckmann) Homotopy groups of maps and exact sequences, Comment. Math. Helv., 34 (1960), 271-304.

[45] Note on the Jacobi identity for Whitehead products, Proc. Cambridge Phil. Soc., 57 (1961), 180-182.

[46] Memorial tribute to J.H.C. Whitehead, Enseignement Mathématique, 7 (1961), 107-125.

[47] On excision and principal fibrations, Comment. Math. Helv., 35 (1961), 77-84.

[48] (with B. Eckmann) Structure maps in group theory, Fund. Math., 50 (1961), 207-221.

[49] (with D. Rees) Natural maps of extension functors and a theorem of R.G. Swan, Proc. Cambridge Phil. Soc., 57 (1961), 489-502.

[50] Note on free and direct products in general categories, Bull. Soc. Math. Belg., 13 (1961), 38-49.

[51] (with B. Eckmann) Group-like structures in general categories I. Multiplications and comultiplications, Math. Ann. 145 (1962), 227-255.

[52] Fundamental group as a functor, Bull. Soc. Math. Belg., 14 (1962), 153-177.

[53] Note on a theorem of Stasheff, Bull.Acad. Polon. Sci., 10 (1962), 127-131.

[54] (with T. Ganea and F.P. Peterson) On the homotopy-commutativity of loop-spaces and suspensions, Topology, 1 (1962), 133-141.

[55] (with B. Eckmann and T. Ganea) On means in general categories, Studies in mathematical analysis and related topics, Stanford (1962), 82-92.

[56] (with B. Eckmann) Group-like structures in general categories II. Equalizers, limits, length, Math. Ann., 151 (1963), 150-186.

[57] (with B. Eckmann) Group-like structures in general categories III. Primitive categories, Math. Ann., 150 (1963), 165-187.

[58] (with S.M. Yahya) Unique divisibility in abelian groups, Acta Math., 14 (1963), 229-239.

[59] Natural group structures in homotopy theory, Lucrarile Consfatuirii de Geometrie so Topologie, Iasi, (1962), 33-37.

[60] Nilpotency and higher Whitehead products, Proc. Coll. Alg. Top., Aarhus (1962), 28-31.

[61] (with B. Eckmann) A natural transformation in homotopy theory and a theorem of G.W. Whitehead, Math. Z., 82 (1963), 115-124.

[62] Remark on loop spaces, Proc. Amer. Math. Soc., 15 (1964), 596-600.

[63] Nilpotency and H-spaces, Topology, 3 (1964), suppl. 2, 161-176.

[64] (with I. Berstein) Suspensions and comultiplications, Topology, 2 (1963), 63-82.

[65] (with I. Berstein) Homomorphisms of homotopy structures, Topology et Géom. Diff. (Sém. C. Ehresmann), Paris (1963), 1-24.

[66] (with B. Eckmann) Unions and intersections in homotopy categories, Comment. Math. Helv., 38 (1964), 293-307.

[67] Spectral sequences for a factorization of maps, Seattle Conference (1963).

[68] (with B. Eckmann) Composition functors and spectral sequences, Comment. Math. Helv., 41 (1966-67), 187-221.

[69] Note on H-spaces and nilpotency, Bull. Acad. Polon. Sci. 11 (1963), 505-509.

[70] (with B. Eckmann) Exact couples in abelian categories, J. Algebra, 3 (1966), 38-87.

[71] Catégories non-abéliennes, Sém. Math. Sup., Université de Montréal (1964).

[72] Exact couples for iterated fibrations, Centre Belge de Rech. Math., Louvain (1966), 119-136.

[73] Correspondences and exact squares, Proc. Conf. Cat. Alg., La Jolla, Springer (1966), 254-271.

[74] (with B. Eckmann) Filtrations, associated graded objects and completions, Math. Z., 98 (1967), 319-354.

[75] Review of 'Modern algebraic topology' by D.G. Bourgin, Bull. Amer. Math. Soc. 71 (1965), 843-850.

[76] The continuing work of CCSM, Arithmetic Teacher (1966), 145-149.

[77*] Homotopy theory and duality, Gordon and Breach (1965).

[78] (with I. Pressman) A generalization of certain homological functors, Ann. Mat. Pura Appl. (4), 71 (1966), 331-350.

[79] On the homotopy type of compact polyhedra , Fund. Math., 61 (1967), 105-109.

[80] On systems of interlocking exact sequences, Fund. Math., 61 (1967), 111-119.

[81*] Studies in modern topology, Introduction, Math. Ass. Amer., Prentice-Hall (1968).

[82] The Grothendieck group of compact polyhedra, Fund. Math., 61 (1967), 199-214.

[83] Arts and Sciences, Methuen, (1967), 20-46.

[84] (with A. Deleanu) Some remarks on general cohomology theories, Math. Scand., 22(1968), 227-240.

[85] Some remarks concerning the semi-ring of polyhedra, Bull. Soc. Math. Belg., 19 (1967), 277-288.

[86] (with B. Eckmann) Commuting limits with colimits, J. Alg., 11 (1969), 116-144.

[87] (with B. Eckmann) Homotopical obstruction theory, An. Acad. Brasil. Ciênc. 40 (1968), 407-429.

[88] Filtrations, Cahiers Topologie Géom. Différentielle, 9 (1967), 243-253.

[89] On the construction of cohomology theories, Rend. Mat. (6), 1 (1968), 219-232.

[90] On commuting limits, Cahiers Topologie Géom. Différentielle, 10 (1968), 127-138.

[91] Categories and functors, Probe, (1968), 33-37.

[92] Note on the homotopy type of mapping cones, Comm. Pure Appl. Math. 21 (1968), 515-519.

[93] (with J. Roitberg) Note on principal S^3-bundles, Bull. Amer. Math. Soc. 74 (1968), 957-959.

[94] (with I. Pressman) On completing bicartesian squares, Proc. Symp. Pure Math., Vol. XVII, Americ. Math. Soc.(1970), 37-49.

[95*] (with H.B. Griffiths) Classical mathematics, Van Nostrand Reinhold (1970).

[96] (with B. Steer) On fibred categories and cohomology, Topology, 8 (1969), 243-251.

[97] (with R. Douglas and F. Sigrist) H-spaces, Springer Lecture Notes, 92 (1969), 65-74.

[98] (with J. Roitberg) On principal S^3-bundles over spheres, Ann. Math., 90 (1969), 91-107.

[99] On factorization of manifolds, Proc. 15th Scand. Math. Congress, Oslo, Springer Lecture Notes 118 (1970), 48-57.

[100] (with Y.C. Wu) On the addition of relations in an abelian category, Canad. J. Math., 22 (1970), 66-74.

[101] On the Ditchley conference and curricular reform, Amer. Math. Monthly, 75 (1969), 1005-1006.

[102*] Algebraic Topology, Courant Institute of Math. Sciences, NYU (1969).

[103*] General cohomology theory and K-theory, Lond. Math. Soc., Lecture Note Series 1, Cambridge University Press (1971).

[104] (with A. Deleanu) On the general Čech construction of cohomology theories, Symposia Mathematica, lst. Naz. di Alt. Mat., (1970), 193-218.

[105] (Ed.) Proc. Battelle Conf. on category theory, homology theory and their applications, Springer Lecture Notes, Vols. 86, 92, 99 (1969).

[106] (with H. Hanisch und W.M. Hirsch) Algebraic and combinatorial aspects of coherent structures, Trans. N.Y. Acad. Sci. II, 31 (1969), 1024-1037.

[107] (with R. Long and N. Meltzer) Research in mathematics education, Educational Studies in Mathematics, 2 (1970), 446-468.

[108] (with A. Deleanu) On the generalized Čech construction of cohomology theories, Battelle Institute Report No. 28, Geneva (1969), 35.

[109] On the category of direct systems and functors on groups, Battelle Institute Report No. 32, Geneva (1970), 40.

[110] Putting coefficients into a cohomology theory, Battelle Institute Report No. 33, Geneva (1970), 34.

[111] (with A. Deleanu) On Kan extensions of cohomology theories and Serre classes of groups, Battelle Institute Report No. 34, Geneva (1970), 40.

[112] (with J. Roitberg) On the classification problem for H-spaces of rank 2, Comment. Math. Helv., 45 (1970), 506-516.

[113] Kategorien, Funktoren und natürliche Transformationen, Math. Phys. Semesterberichte, 17 (1970), 1-12.

[114] Cancellation: A Crucial Property? New York State Math. Teachers Journal, 20 (1970), 68-74, 132-135.

[115] The Cosrims Reports, Amer. Math. Monthly, 77 (1970), 515-517.

[116] Putting coefficients into a cohomology theory, Proc. Koninkl. Nederl. Akad. van Wetenschappen, Series A, 73 (1970), 196-216.

[117] On the category of direct systems and functors on groups, Pure Appl. Alg., 1 (1971), 1-26.

[118] (with A. Deleanu) On Kan extensions of cohomology theories and Serre classes of groups, Fund. Math., 73 (1971), 143-165.

[119] (with J. Roitberg) Note on quasifibrations and fibre bundles, Ill. J. Math., 15 (1971), 1-8.

[120] (with G. Mislin and J. Roitberg) Sphere bundles over spheres and non-cancellation phenomena, Springer Lecture Notes 249 (1971) 34-46.

[121*] Lectures on Homological Algebra, Reg. Conf. Series in Math., Amer. Math. Soc. (1971).

[122] (with B. Eckmann) On central extensions and homology, Battelle Institute Report No. 45, Geneva (1971), 18.

[123] On filtered systems of groups, colimits, and Kan extensions, J. Pure Appl. Alg., 1 (1971), 199-217.

[124] Extensions of functors on groups and coefficients in a cohomology theory, Actes Congrès intern. math., 1970, Tome 1, 1-6.

[125] Topology in the high school, Educational Studies in Mathematics 3 (1971), 436-453.

[126] The role of categorical language in pre-college mathematical education.

[127] Should mathematical logic be taught formally in mathematics classes? The Mathematics Teacher, 64 (1971), 389-394.

[128] Direkte Systeme von Gruppen, Math. Phys. Semesterberichte, 18 (1971), 174-193.

[129] (with J. Roitberg) On the classification of torsion-free H-spaces of rank 2, Springer Lecture Notes 168 (1970), 67-74.

[130] (with A. Deleanu) Remark on Čech extension of cohomology functors, Proc. Adv. Stu. Inst. Alg. Top., Aarhus (1970), 44-66.

[131] (with A. Deleanu) On the splitting of universal coefficient sequences, Proc. Adv. Stu. Inst. Alg. Top., Aarhus (1970), 180-201.

[132] On direct systems of groups, Bol. Soc. Brasil. Mat. 2 (1970), 1-20.

[133] (with J. Roitberg) On quasifibrations and orthogonal bundles, Springer Lecture Notes 196 (1971), 100-106.

[134] (with B. Eckmann and U. Stammbach) On Ganea's extended homology sequence and free presentations of group extensions, Battelle Institute Report No. 48, Geneva (1971), 23.

[135] (with A. Deleanu) Localization, homology and a construction of Adams, Battelle Institute Report No. 47 (1971), 31.

[136] (with B. Eckmann and U. Stammbach) On the homology theory of central group extensions, Comment. Math. Helv., 47 (1972), 102-122.

[137] Some problems of contemporary university education, Bol. Soc. Brasil Mat., 2 (1971), 67-75.

[138] Non-cancellation phenomena in topology, Coll. Math. Soc. Janos Bolyai 8. Topics in Topology, Keszthely, Hungary (1972), 405-416.

[139] (with B. Eckmann and U. Stammbach) On the homology theory of central group extensions II. The exact sequence in the general case, Comment. Math. Helv. 47 (1972), 171-178.

[140] (with B. Eckmann) On central group extensions and homology, Comment. Math. Helv. 46 (1971), 345-355.

[141*] Topicos de algebra homológica, Publicaçoes do Instituto de Matematica e Estatistica, Universidade de São Paulo (1970).

[142*] (with U. Stammbach) A course in homological algebra, Springer (1971).

[143] Topologie à l'école secondaire, Niko 8 (1971), 2-28; (in dutch) 69-95.

[144] Suites spectrales et théories de cohomologie générales, Université de Montpellier, Publication No. 88 (1969-1970).

[145] Mélanges d'algèbre pure et appliquée, Université de Montpellier, Publication No. 83 (1969-1970).

[146] The art of mathematics, Kynoch Press, Birmingham, 1960, 14.

[147] Arithmetic - down but not out Teaching Arithmetic, (1963), 9-13.

[148] Arithmetic as part of mathematics, Mathematics Teaching, 27 (1964).

[149] Categorieën en functoren, Niko 9 (1971), 85-100.

[150] (with G. Mislin and J. Roitberg) Note on a criterion of Scheerer, Ill. Journ. Math. 17 (1973), 680-688.

[151] Obituary: Heinz Hopf, Bull. London Math. Soc. 4 (1972), 202-217.

[152] (with G. Mislin and J. Roitberg) H-Spaces of rank two and non-cancellation phenomena, Invent. Math. 16 (1972), 325-334.

[153] Categories and homological algebra, Encyclopaedia Britannica, (1974), 547-554.

[154] Extensions of functors on groups and coefficients in a cohomology theory, Actes du Congrès International des Mathématiciens, Nice 1970, Tome 1, 323-328.

[155] (with B. Eckmann and U. Stammbach) On the Schur multiplicator of a central quotient of a direct product of groups, J. Pure Appl. Algebra, 3 (1973), 73-82.

[156] (with G. Mislin and J. Roitberg) Topological localization and nilpotent groups, Bull. Amer. Math. Soc. 78 (1972), 1060-1063.

[157] (with G. Mislin and J. Roitberg) Homotopical localization, Proc. London Math. Soc. (3), 24 (1973), 693-706.

[158] (with A. Deleanu) Localization, homology and a construction of Adams, Trans Amer. Math. Soc. 179 (1973), 349-362.

[159] The language of categories in high school mathematics, Niko 15 (1973), 17-55.

[160] A mathematician's miscellany, Battelle Seattle Research Center Lecture Series 72-1 (1972).

[161] (with G. Mislin and J. Roitberg) Sphere bundles over spheres and non-cancellation phenomena. London Math. Soc. 6 (1972), 15-23.

[162] (with U. Stammbach) Two remarks on the homology of group extensions, J. Austral. Math. Soc., 17 (1974), 345-357.

[163*] Category theory, NSF Short Course, Colgate University (1972).

[164] In the bag, New York State Mathematics Teachers' Journal, 23 (1973), 8-11.

[165] (with U. Stammbach) On torsion in the differentials of the Lyndon-Hochschild-Serre spectral sequence, J. Algebra 29 (1974), 349-367.

[166] (with U. Stammbach) On the differentials of the Lyndon-Hochschild-Serre spectral sequence, Bull. Amer. Math. Soc. 79 (1973), 796-799.

[167] Localization and cohomology of nilpotent groups, Math. Z. 132 (1973), 263-286.

[168] (with A. Deleanu and A. Frei) Generalized Adams completion, Cahiers Topologie Géom. Différentielle 15 (1974), 61-82.

[169] Remarks on the localization of nilpotent groups, Comm. Pure Appl. Math. 24 (1973), 703-713.

[170] Some problems of contemporary education, Papers on Educational Reform, Vol. 4 (1974), Open Court Publishing Company, 77-104.

[171*] (with Y.-C. Wu) A course in modern algebra, John Wiley and Sons, 1974.

[172] Localization of nilpotent spaces, Springer Lecture Notes 428 (1974), 18-43.

[173] The survival of education, Educational Technology (1973).

[174] (with A. Deleanu and A. Frei) Idempotent triples and completion, Math. Z. 143 (1975), 91-104.

[175] Localization in topology, Amer. Math. Monthly, 82 (1975), 113-131.

[176*] Le langage des catégories, Collection Formation des Maîtres, Cedic. Paris (1973).

[177] The category of nilpotent groups and localization, Colloque sur l'algèbre des catégories, Amiens, 1973, Cahiers Topologie Géom. Différentielle 14 (1973), 31-33.

[178] Nilpotent actions on nilpotent groups, Springer Lecture Notes, 450 (1975), 174-196.

[179] On direct limits of nilpotent groups, Springer Lecture Notes 418 (1974), 68-77.

[180] The training of mathematicians today, Mathematiker über die Mathematik, edited by Michael Otte, Springer Verlag (1974).

[181] A case against managerial principles in education.

[182] Ten years with Springer Verlag, Springer Verlag, New York (1974).

[183] Localization of nilpotent groups: homological and combinatorial methods, Comptes Rendus des Journées Mathématiques S.M.F., Montpellier, (1974), 123-132.

[184] Education in mathematics and science today; the spread of false dichotomies, Proc. 3rd Int. Congr. Math. Ed., Karlsruhe (1976), 75-97.

[185*] Homologie des groupes, Collection Mathématique, Université Laval, (1973).

[186*] (with G. Mislin and J. Roitberg) Localization of nilpotent groups and spaces. North Holland (1975).

[187] (with G. Mislin) Bicartesian squares of nilpotent groups, Comment. Math. Helv. 50 (1975), 477-491.

[188] (with G. Mislin) On the genus of a nilpotent group with finite commutator subgroup, Math. Z. 146 (1976), 201-211.

[189] (with U. Stammbach) On group actions on groups and associated series, Math. Proc. Cambridge Philos. Soc. 80 (1976), 43-55.

[190] (with A. Deleanu) On the categorical shape of a functor, Fund. Math. 97 (1977) 157-176.

[191] (with G. Mislin) Remarkable squares of homotopy types, Bol. Soc. Brasil. Mat. 5 (1974), 165-180.

[192] (with A. Deleanu) Borsuk shape and a generalization of Grothendieck's definition of pro-category. Math. Proc. Cambridge Philos. Soc. 79 (1976), 473-482.

[193] The new emphasis on applied mathematics, Newsletter, Conference Board of the Mathematical Sciences, vol. 10, 2 (1975), 17-19.

[194] What is modern mathematics?, Pokroky Matematiky Fyziky a Astronomie, 3 (1977), 151-164.

[195] (with J. Roitberg) Generalized C-theory and torsion phenomena in nilpotent spaces. Houston J. Math. 2 (1976), 525-559.

[196] (with U. Stammbach) Localization and isolators, Houston J. Math. 2 (1976), 195-206.

[197] A humanist's assault on managerial principles in education.

[198] Localization of nilpotent spaces. Lecture Notes in Pure and Applied Mathematics, vol. 12, Marcel Dekker (1975), 75-100.

[199] The anatomy of a conference.

[200] (with J. Roitberg) On the Zeeman comparison theorem for the homology of quasi-nilpotent fibrations, Quart. J. Math. Oxford (2), 27 (1976), 433-444.

[201] (with A. Deleanu) On Postnikov-true families of complexes and the Adams completion, Fund. Math. 106 (1980), 53-65.

[202] On Serre classes of nilpotent groups, Proc. Conference in Honor of Candido do Lima da Silva Dias.

[203] Unfolding of singularities, Proceedings of the São Paulo Symposium on Functional Analysis, Lecture Notes in Pure and Applied Mathematics, vol. 18, Marcel Dekker (1976), 111-134.

[204] What experiences should be provided in graduate school to prepare the college mathematics teacher?, The Bicentennial Tribute to American Mathematics, Math. Ass. Amer. (1977), 187-191.

[205] (with A. Deleanu) Note on homology and cohomology with $\mathbb{Z}_p$-coefficients, Czechoslovak Math. J. 28 (103), 474-479.

[206] (with A. Deleanu) On a generalization of Artin-Mazur completion (preprint).

[207] (with R. Fisk) Derivatives without limits (preprint).

[208] Basic mathematical skills and learning: Position Paper, NIE Conference, vol. 1 (1976), 88-92.

[209] (with G. Rising) Thoughts on the state of mathematics education today, NIE Conference, vol. 2 (1976), 33-42.

[210] Structural stability, catastrophe theory and their applications in the sciences, Research Futures, Battelle Institute (1976).

[211] Localization and cohomology of nilpotent groups, Cahiers Topologie Géom. Différentielle, 14 (1973), 341-370.

[212] Localization in group theory and homotopy theory, Jber. Deutsch. Math.-Verein, 79 (1977), 70-78.

[213] (with David Singer) On G-nilpotency (preprint).

[214] (with G. Mislin and J. Roitberg) On maps of finite complexes into nilpotent spaces of finite type: a correction to 'Homotopical localization', Proc. London Math. Soc. 34 (1978), 213-225.

[215] On G-spaces, Bol. Soc. Brasil. Mat. 7 (1976), 65-73.

[216] (with A. Deleanu) Generalized shape theory, Springer Lecture Notes 609 (1977), 56-65.

[217] Localization theories for groups and homotopy types, Springer Lecture Notes 597 (1977), 319-329.

[218] (with G. Mislin and J. Roitberg) On co-H-spaces, Comment. Math. Helv. 53 (1978) 1-14.

[219] Some contributions of Beno Eckmann to the development of topology and related fields, L'Enseignement Mathématique, 23 (1977), 191-207.

[220] (with Joseph Roitberg) On the finitude of counterimages in maps of function-spaces: Correction to 'Generalized C-theory and torsion phenomena in nilpotent spaces', Houston J. Math. 3 (1977) 235-238.

[221] Nilpotency in group theory and topology,

[222] A Friendship and a Bond; Semper Attentus, Beiträge für Heinz Götze zum 8. August 1977, Springer Verlag (1977), 150-153.

[223] (with G. Goldhaber), NAS-NRC Committee on Applied Mathematics Training, Notices Amer. Math. Soc. (1977), 435-436.

[224] The changing face of mathematics, Case Alumnus, 57, 2(1977), 5-7.

[225] (with C. Cassidy) L'isolateur d'un homomorphisme de groupes, Canad. J. Math. 31 (1979), 375-391.

[226] (with G. Mislin, J. Roitberg and R. Steiner) On free maps and free homotopies into nilpotent spaces, Springer Lecture Notes 673 (1978), 202-218.

[227] On orbit sets for group actions and localization, Springer Lecture Notes 673 (1978), 185-201.

[228] (with J. Roitberg) On a generalization of two exact sequences of Steiner, Ill. J. Math. 24 (1980), 206-215.

[229] Teaching and Research: A false dichotomy, Mathematical Intelligencer, 1 (1978), 76-80.

[230] (with Joseph Roitberg and David Singer) On G-spaces, Serre classes, and G-nilpotency, Math. Proc. Cambridge Philos. Soc. 84 (1978), 443-454.

[231] Vector spaces, abelian groups and groups: similarities and differences, Atas do 11º Coloquis da SBM, 1977.

[232] (with Carl Bereiter and Stephen Willoughby) Real Math, Grade 1, Open Court Publishing Company (1978).

[233] Some thoughts on math anxiety, I. Thoughts on diagnosis, Ontario Mathematics Gazette, 17, 1 (1978), 35-42.

[234] Some thoughts on math anxiety, II. Thoughts on cure, Ontario Mathematics Gazette, 17, 2 (1978), 26-28.

[235] (with Carl Bereiter and Stephen Willoughby) Real Math, Grade 2, Open Court Publishing Company (1978).

[236] (with Joseph Roitberg) Profinite completion and generalizations of a theorem of Blackburn, J. Algebra 60 (1979), 289-306.

[237] Dangerous division, California Mathematics, 4 (1979), 15-21.

[238] (with Paulo Leite) On nilpotent spaces and C-theory, C.R. Math. Rep. Acad. Sci. Canada, 1 (1979), 125-128.

[239] Review of 'Obstruction theory' by Hans J. Baues, Bull. Amer. Math. Soc. 1 (1979), 292-398.

[240] Review of 'Why the professor can't teach' by Morris Kline, Amer. Math. Monthly, 86 (1979), 407-412.

[241] (with Carl Bereiter and Stephen Willoughby) Real Math, Grade 3, Open Court Publishing Company (1979).

[242] Duality in homotopy theory: a retrospective survey. J. Pure Appl. Algebra, 19 (1980), 159-169.

[243] The role of applications in the undergraduate mathematics curriculum, Ad Hoc Committee. Applied Mathematics Training (Chairman, Peter Hilton), NRC (1979), 25.

[244] Math anxiety; some suggested causes and cures, part I, Two-Year-College Mathematics Journal 11 (1980), 174-188.

[245] (with Carl Bereiter, Joseph Rubinstein and Stephen Willoughby) Real Math, Grade 4, Open Court Publishing Company (1980).

[246] (with Jean Pedersen) Review of 'Overcoming math anxiety' by Sheila Tobias, and 'Mind over math' by Stanley Kogelman and Joseph Warren, Amer. Math. Monthly 87 (1980), 143-148.

[247*] (with Jean Pedersen) Fear no more: An adult approach to mathematics, Addison Wesley (1982).

[248] Math anxiety: Some suggested causes and cures, Part II, Two-Year-College Mathematics Journal 11 (1980), 246-251.

[249] (with Jean Pedersen) Teaching mathematics to adults.

[250] Do we still need to teach fractions, Proc. ICME 4.

[251] (with Joseph Roitberg) Restoration of structure, Cahiers Topologie Géom. Différentielle, 22 (1981), 201-207.

[252] (with Jean Pedersen) On the distribution of the sum of a pair of integers.

[253] (with Jean Pedersen) Casting out 9's revisited, Math Mag. 54 (1981), 195-201.

[254] The emphasis on applied mathematics today and its implications for the mathematics curriculum, New Directions in Applied Mathematics, Springer (1982), 155-163.

[255] Avoiding math avoidance, Mathematics Tomorrow, Springer Verlag (1981), 73-83.

[256] (with Joseph Roitberg) Note on completions in homotopy theory and group theory.

[257] (with Carl Bereiter, Joseph Rubinstein and Steve Willoughby) I. How deep is the water? II. Measuring bowser. III. Bargains galore, Thinking stories, Open Court Publishing Company, (1981), 91, 121, 121 (1981).

[258] (with Carl Bereiter, Joseph Rubinstein and Steve Willoughby) Real Math, Grade 5, Open Court Publishing Company (1981), 511.

[259] (with Carl Bereiter, Joseph Rubinstein and Steve Willoughby) Real Math, Grade 6, Open Court Publishing Company (1981), 425.

[260] Group structure and enriched structure in homotopy theory, Proc. Math. Sem. Singapore (1980), 31-38.

[261] The language of categories and category theory, Math. Intelligencer 3 (1981), 79-82.

[262] Group structure in homotopy theory and generalizations of a theorem of Blackburn, Atas.

[263] Relative nilpotent groups, Proc. Carleton Conference on Algebra and Topology (1981).

[264] Some trends in the teaching of algebra, Proc. Hongkong Conference on New Trends in Mathematics (1981).

[265] The education of applied mathematics, SIAM News 14, 5, October (1981).

[266] Homotopy, Encyclopedia Britannica.

[267] (with Jean Pedersen) $e^{\pi} > \pi^{e}$?, Mathematics Teacher 74 (1981), 501-502.

[268] Nilpotent groups and abelianization, Questiones Mathematicae (1982).

[269] Groupes relatifs et espaces relatifs, Proc., 6ième Congrès du Groupement des Mathématiciens d'Expression Latine (1982).

[270] Review of 'Mathematics: The lose of certainty" by Morris Kline, Bull. London Math. Soc. (1982).

[271] Reflections on a visit to South Africa, Focus, Math. Ass. Amer., November, (1981).

[272] ICMI, 1980-81, L'Enseignement Mathématique (1982).

[273] Mathematics in 2001 - Implications for today's Undergraduate Teaching, Proc. Conf. Remedial and Developmental Mathematics in Colleges, New York, April, 1981.

[274] (with Gail Young, ed) New directions in applied mathematics, Springer (1982).

[275] Review 'A Brief course of higher mathematics, by V.A. Kudiyovtsev and B.P. Demidavich, Amer. Math. Monthly (1982).

ESSAY ON HILTON'S WORK IN TOPOLOGY

Guido Mislin, ETH Zürich

INTRODUCTION

Peter Hilton's work in topology covers a wide range of topics, some have a distinct geometric character, others are more algebraic. Often a basic idea is modulated through different keys, creating new variations of old concepts in a different category. A typical example is that of the topological notion of homotopy, which is transformed into the algebraic concept of projective and injective homotopy for modules. The resulting algebraic structures were studied jointly by Beno Eckmann and Peter Hilton. Passing back from algebra to topology led to the origin of the Eckmann-Hilton duality (cf. Urs Stammbach's article in this volume). Hilton's early work concerned mainly homotopy groups, Hopf invariants, Whitehead products and the homotopy groups of a wedge of spaces (Hilton-Milnor formula). In the first section of this essay we will concentrate on Hilton's paper "On the homotopy groups of the union of spheres" [H13], which is the perfect example of his clarity of exposition and his ability to use efficiently the interplay of algebra and geometry. The topics which follow are not meant to represent his entire work. They serve only as illustrations of different aspects of his mathematical accomplishments.

The following table illustrates the main areas of Peter Hilton's work in topology.

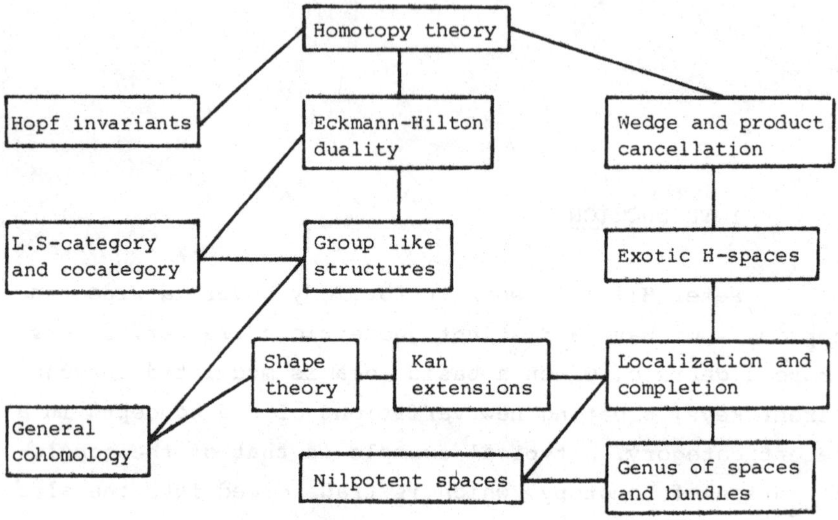

Peter Hilton is not only a fine research mathematician, he is a great teacher and mentor. Of his more than thirty coauthors, many were young and inexperienced when they first worked with him; these beneficial professional collaborations have often lead to lifelong friendships. I happily count myself one of these fortunate people. Thank you Peter.

1. THE HILTON-MILNOR FORMULA

A homotopy operation in k variables is a natural transformation

$$\pi_{n_1}(X) \times \dots \times \pi_{n_k}(X) \to \pi_n(X)$$

Such operations correspond bijectively to the elements of $\pi_n(S^{n_1} \vee \ldots \vee S^{n_k}) = \pi_n(T)$; $T := S^{n_1} \vee \ldots \vee S^{n_k}$. Hilton computed $\pi_n(T)$ for a 1-connected T in [H13], expressing it in terms of homotopy groups of spheres. His calculation makes use of the relationship between $\pi_*(T)$ and $H_*(\Omega T)$, the Whitehead product in $\pi_*(T)$ and the Pontryagin product in $H_*(\Omega T)$. The following is a short outline of his beautiful proof.

By a result of Bott and Samelson [2] the Pontryagin algebra $H_*(\Omega T) = R$ is a free associative algebra, freely generated by elements $e_1,\ldots,e_k$, which correspond to the spheres $S^{n_1},\ldots S^{n_k}$ via transgression: thus R is a coproduct of rings $\mathbb{Z}[e_1] * \ldots * \mathbb{Z}[e_k]$, the generator e_j having degree n_j-1 . A theorem of Samelson [16] implies that the composite map

$$\rho : \pi_n(T) \overset{\cong}{\to} \pi_{n-1}(\Omega T) \overset{Hu}{\to} H_{n-1}(\Omega T)$$

fulfills $\rho[\alpha,\beta] = (-1)^p(\rho(\alpha)\rho(\beta) - (-1)^{pq}\rho(\beta)\rho(\alpha))$

for $\alpha \in \pi_{p+1}(T)$ and $\beta \in \pi_{q+1}(T)$. This lead Hilton to define the quasi-commutator $[a,b]$ for two elements $a,b \in R$ of gradation p and q by setting

$$[a,b] = (-1)^p(ab-(-1)^{pq}ba)$$

The map $\rho : \pi_*(T) \to H_*(\Omega T)$ of degree -1 then fulfills $\rho[\alpha,\beta] = [\rho(\alpha),\rho(\beta)]$.
The additive structure of $R = \mathbb{Z}[e_1] * \ldots * \mathbb{Z}[e_k]$ can be described as follows. By an inductive procedure, basic products in R are defined and ordered. The basic products of weight 1 are the generators $e_1,\ldots,e_k \in R$; they are ordered by setting $e_1 < e_2 < \ldots < e_k$. If the basic products of weight $< r$ are defined and ordered, then the basic products of weight r are defined to be the elements $[a,b] \in R$ where a and b are basic products of weight u and v respectively,

$u + v = r$, $a < b$, and if $b = [c,d]$, then one must have $c \leq a$. The basic products of weight r are given an arbitrary order and they are considered greater than basic products of lesser weight. If $z_1,\dots z_m \in R$ are basic products, the monomial $z_1 \dots z_m$ is said to have zero disorder, if $i < j$ ($\leq m$) implies $z_i \leq z_j$. Hilton proved the following crucial result, generalizing ideas of M. Hall [6], Ph. Hall [7] and Magnus [13].

<u>Theorem 1:</u> The monomials of zero disorder in basic products of elements of $R = \mathbb{Z}[e_1] * \dots * \mathbb{Z}[e_k]$ form an additive basis for R .

For each basic product $w \in R$ an element $w(\pi) \in \pi_*(T)$ is defined by replacing each e_j by the inclusion class $\iota_j : S^{n_j} \to T$ and the bracket in R by the Whitehead product in $\pi_*(T)$; these elements $w(\pi)$ are called basic Whitehead products. Note that $\rho(w(\pi)) = w$. Thus, if $w(\pi) : S^{n_w} \to T$, the map $\Omega(w(\pi)) : \Omega S^{n_w} \to \Omega T$ maps the canonical generator x of the polynomial ring $H_*(\Omega S^{n_w}) = \mathbb{Z}[x]$ to $w \in H_*(\Omega T)$. It follows then in view of Theorem 1 that the product map

$$\tilde{\prod_w} \Omega S^{n_w} \to \Omega T$$

induces a homology equivalence; it is therefore a homotopy equivalence, because the spaces in question are H-spaces. Hilton's theorem can now be stated as follows.

<u>Theorem 2:</u> $\pi_n(S^{n_1} \vee \dots \vee S^{n_k}) \cong \bigoplus_{w(\pi)} \pi_n(S^{n_w})$

where $w(\pi) \in \pi_*(S^{n_1} \vee \dots \vee S^{n_k})$ runs over all basic Whitehead products and where the direct summand $\pi_n(S^{n_w})$ is embedded by composition with $w(\pi) : S^{n_w} \to S^{n_1} v \dots v S^{n_k}$.

The theorem may be used to give a short proof for the "Jacobi Identity" for Whitehead products. Setting $L_n(X) = \pi_{n+1}(X)$ and using the Whitehead product as multiplication, one obtaines a graded ring $L_*(X)$ for any 1-connected space X . This ring $L_*(X)$ is a quasi-Lie ring in the sense of Hilton. He studied the structure of quasi-Lie rings in general and determined the additive structure of free quasi-Lie rings in [H20]. It turns out that the only relations which hold in $L_*(X)$ for all X , are the relations which hold in a free quasi-Lie ring.

Theorem 2 was generalized by Milnor in the following way (cf. Adam's student guide [1]).

Hilton-Milnor Formula

Let $X_1, \dots X_k$ be connected CW-complexes. Then $\Omega\Sigma(X_1 \vee X_2 \vee \dots \vee X_k)$ has the same homotopy type as a weak infinite product $\tilde{\pi}_{j \geqslant 1} \Omega\Sigma X_j$, where each X_j , $j > k$, is an iterated reduced join $X_1^{(n_1)} \wedge \dots \wedge X_k^{(n_k)}$. The number of factors of a given form is equal to

$$\frac{1}{n} \sum_{d/\delta} \frac{\mu(d)(n/d)!}{(n_1/d)! \dots (n_k/d)!} \quad \text{where} \quad n = n_1 + \dots + n_k ,$$

$\delta = \gcd(n_1, \dots, n_k)$ and μ denotes the Möbius function.

The following application is contained in Hilton's papers [H79] and [H82]. Let $\Sigma\mathbb{P}^1$ denote the set of homotopy types of 1-connected finite polyhedra, which are homotopy equivalent to suspensions. Consider $\Sigma\mathbb{P}^1$ as a monoid, using the wedge operation, and denote by $G(\Sigma\mathbb{P}^1)$ the associated Grothendieck group; write $[X]$ for the element represented by X in $G(\Sigma\mathbb{P}^1)$.

Theorem 3: Let X and Y be 1-connected polyhedra, which are of the homotopy type of suspension spaces. If $[X] - [Y] \in G(\Sigma\mathbb{P}^1)$ is a torsion element, then X and Y have isomorphic homotopy groups.

For example, if X,Y and A are 1-connected finite suspension spaces such that $X \vee A \simeq Y \vee A$, then it follows that $\pi_i(X) \cong \pi_i(Y)$ for all i, since $[X] = [Y]$ in this case; a generalization of this result is due to Kozma [12], a student of Hilton. Note that it is still an open question whether $G(\Sigma\mathbb{P}^1)$ possesses any non-trivial torsion elements.

2. THE HILTON-HOPF INVARIANTS

The classical Hopf homomorphism $H : \pi_{2r-1}(S^r) \to \mathbb{Z}$ (cf. [8]) was generalized first by G.W. Whitehead to a homomorphism $H : \pi_n(S^r) \to \pi_n(S^{2r-1})$, $n < 3r-3$, and later, by Hilton to groups with $n \leqslant 4r-4$ (cf. [H3]). Hilton's definition is as follows. One considers the map

$$H^* : \pi_n(S^r) \xrightarrow{\Phi} \pi_n(S^r \vee S^r) \xrightarrow{Q} \pi_{n+1}(S^r \times S^r, S^r \vee S^r) \xrightarrow{\chi} \pi_{n+1}(S^{2r})$$

where Φ is induced by pinching the equator of S^r to a point, Q is the projection onto a direct summand and χ is induced by shrinking $S^r \vee S^r$ to a point. Note that the Freudenthal suspension $E : \pi_n(S^{2r-1}) \to \pi_{n+1}(S^{2r})$ is an isomorphism for $n \leqslant 4r-4$. The map H is then given by

$$H = E^{-1} \circ H^* : \pi_n(S^r) \to \pi_n(S^{2r-1}) \text{ , } n \leqslant 4r-4 \text{ .}$$

In [H13] Hilton proposed the following natural generalization of the homomorphism H. Recall that (Theorem 2)

$$\pi_n(S^r \vee S^r) = \pi_n(S^r) \oplus \pi_n(S^r) \oplus \pi_n(S^{2r-1}) \oplus (\pi_n(S^{3r-2}) \oplus \pi_n(S^{3r-2})) \oplus \ldots$$

where the right hand summands are embedded in $\pi_n(S^r \vee S^r)$ by composition with basic Whitehead products. Let H'_{i-1} $i = 1,2,\ldots,$ denote the projection onto the $(i+2)$-nd summand and set $H_{i-1} = H'_{i-1} \circ \Phi$, $\Phi : \pi_n(S^r) \to \pi_n(S^r \vee S^r)$ as above. This defines the Hilton-Hopf homomorphisms

$$H_o \ : \pi_n(S^r) \to \pi_n(S^{2r-1})$$

$$H_1, H_2 \ : \pi_n(S^r) \to \pi_n(S^{3r-2}) \ \ldots$$

Hilton proves, that H_o agrees with the previously defined H , and $E \circ H_o$ equals H^* . The Hilton-Hopf homomorphisms may be used to express the left distributive law for composition of homotopy classes as follows. Let $\gamma \in \pi_n(S^r)$. Then, by definition, one obtains in $\pi_n(S^r \vee S^r)$ the following equation

$$\Phi(\gamma) = \iota_1 \circ \gamma + \iota_2 \circ \gamma + [\iota_1, \iota_2] \circ H_o(\gamma)$$

$$+ [\iota_1, [\iota_1, \iota_2]] \circ H_1(\gamma) + [\iota_2, [\iota_1, \iota_2]] \circ H_2(\gamma) + \ldots$$

where ι_1 and ι_2 denote the homotopy classes of the two inclusions $S^r \to S^r \vee S^r$. Thus, if $\alpha, \beta \in \pi_r(X)$ and $\gamma \in \pi_n(S^r)$,

$$(\alpha+\beta) \circ \gamma = \alpha \circ \gamma + \beta \circ \gamma + [\alpha, \beta] \circ H_o(\gamma)$$

$$+ [\alpha, [\alpha, \beta]] \circ H_1(\gamma) + [\beta, [\alpha, \beta]] \circ H_2(\gamma) + \ldots$$

The Hilton-Hopf invariants measure therefore the deviation from additivity for the homotopy operation

$$\gamma^* : \pi_r(X) \to \pi_n(X) \ , \ \alpha \mapsto \alpha \circ \gamma = \gamma^*(\alpha)$$

For $k \in \mathbb{Z}$ one obtains the simple formula

$$\gamma^*(k\alpha) = k\gamma^*(\alpha) + \binom{k}{2}[\alpha,\alpha] \circ H_0(\gamma) + 2\binom{k+1}{3}[\alpha,[\alpha,\alpha]] \circ H_1(\gamma)$$

and the last term, involving $[\alpha,[\alpha,\alpha]]$, vanishes if r is odd.

3. ECKMANN-HILTON DUALITY

In the Comptes Rendus notes [H23,H24,H25] Eckmann and Hilton presented for the first time a frame work for an internal duality in the pointed homotopy category. Their starting point was the homotopy set $[X,Y]$, which can be considered as a functor in X, or "dually" as a functor in Y. They introduced in a self-dual way homotopy groups $\pi_n(X;Y) \cong [\Sigma^n X,Y] \cong [X,\Omega^n Y]$ generalizing simultaneously cohomology groups and homotopy groups; by relativizing with respect to either variable they obtained dual long exact sequences, and the term "cofibration" appeared, dual to the notion of a fibration. In view of this duality, a pair of spaces should be considered as a map, and a triad as a diagram $X \to Y \to Z$. This point of view leads to dual triple sequences in homotopy and cohomology, which were used by Eckmann and Hilton [H30] to define the homotopy and homology decomposition of a map. In later papers [H51,H56,H57] they studied internal duality in arbitrary categories. In particular, they analysed the concepts of groups and cogroups in a general setting. The reader who wishes to find out more about the history and meaning of duality is referred to Hilton's recently published review article [H242]. We will try to give an idea of Eckmann-Hilton duality by looking from a dualists point of view at a specific example.

To fix ideas, we place ourself in the pointed homotopy category of connected CW-complexes. An H-space X is a space for which the folding map $\nabla : X \vee X \to X$ extends to $X \times X$:

$$\begin{array}{ccc} X \times X & \dashrightarrow & X \\ \uparrow & \nearrow_{\nabla} & \\ X \vee X & & \end{array}$$

Dually, a coH-space Y is a space for which the diagonal map $\Delta : Y \to Y \times Y$ may be compressed into $Y \vee Y$:

$$\begin{array}{ccc} Y \vee Y & \dashleftarrow & Y \\ \downarrow & \swarrow_{\Delta} & \\ Y \times Y & & \end{array}$$

By a result of James [10], X is an H-space if and only if the canonical map $X \to \Omega\Sigma X$ has a left inverse. Dually, Y is a coH-space if and only if the canonical map $\Sigma\Omega Y \to Y$ has a right inverse; this has been proved by Ganea [4]. James proved his theorem using the "James model" for $\Omega\Sigma X$; there is no duality visible relating his proof to Ganea's.

It is well known that every H-structure $\mu : X \times X \to X$ admits an inversion on each side: we say (X,μ) is a loop-object. This may equivalently be expressed by the fact that the two shift maps

$$X \times X \xrightarrow{1\times\Delta} X \times X \times X \xrightarrow{\mu\times 1} X \times X , \text{ and}$$

$$X \times X \xrightarrow{\Delta\times 1} X \times X \times X \xrightarrow{1\times\mu} X \times X$$

are homotopy equivalences for every H-structure μ on X . But not every coH-structure $\nu : Y \to Y \vee Y$ admits inverses, that is, the corresponding maps

$$Y \vee Y \xrightarrow{1\vee\nu} Y \vee Y \vee Y \xrightarrow{\nabla\vee 1} Y \vee Y , \text{ and}$$

$$Y \vee Y \xrightarrow{\nu \vee 1} Y \vee Y \vee Y \xrightarrow{1 \vee \nabla} Y \vee Y$$

need not be homotopy equivalences: there are coH-structures such that (Y,ν) is not a coloop-object (examples are due to Barratt). However, if Y is 1-connected, one can dualize the proof for H-spaces and show that for every coH-structure ν on Y, (Y,ν) is a coloop-object (cf. [H218]).

It is an elementary fact that every finite H-space X is of the form $S \times Z$, where S is a product of circles and $H^1(Z) = 0$. Dually, one might expect the following to hold (cf. [5]).

Ganea Conjecture: Let Y be a finite (connected) coH-space. Then $Y \simeq T \vee W$ with T a wedge of circles and $\pi_1(W) = 0$.

The Ganea conjecture has been proved for coloop-objects [H218]. Note that no coH-spaces Y are known which do not admit a coloop structure $Y \to Y \vee Y$.

4. THE HILTON-ROITBERG MANIFOLD AND LOCALIZATION

Hopf's original paper [9] marked the beginning of the study of topological groups from a homotopy point of view. When, more than 25 years later, Hilton and Roitberg found the first example of a finite Hopf-space different from the obvious ones (Lie groups, S^7, $P_7(\mathbb{R})$ and their products), the study of H-spaces became a very active field of research. The new H-space, which was denoted by $E_{7\omega}$, was shown by Stasheff to be of the homotopy type of a topological group [18]. One can describe $E_{7\omega}$ as follows. Consider the principal S^3-bundle $Sp(1) \hookrightarrow Sp(2) \to S^7$ and let $S^3 \to E_{n\omega} \to S^7$

denote the bundle induced by a map $S^7 \to S^7$ of degree n. Then, as we will sketch below, $E_{7\omega} \not\simeq Sp(2)$ but $E_{7\omega} \times S^3$ and $Sp(2) \times S^3$ are diffeomorphic. This implies that the manifold $E_{7\omega}$ is an H-space, being a retract of the Lie group $Sp(2) \times S^3$.

The basic construction behind this type of example was first considered by Hilton [H82] in a dual setting, where he demonstrated the failure of wedge-cancellation. His dual examples are as follows. Let $\alpha, \beta \in \pi_{m-1}(S^n)$ and let C_α, C_β be the mapping cones of α and β. Using a form of the Blakers-Massey Theorem, it follows that $C_\alpha \simeq C_\beta$ if and only if $\pm\beta = (\pm 1) \circ \alpha$. Suppose now that α has finite order k and that $\beta = \ell\alpha$, with ℓ prime to k. By forming the push-out diagram

$$\begin{array}{ccccc} S^{m-1} & \xrightarrow{\alpha} & S^n & \xrightarrow{\iota_\alpha} & C_\alpha \\ & & {\scriptstyle \iota_\beta}\downarrow & & \downarrow \\ & & C_\beta & \rightarrow & C_{\alpha\beta} \end{array}$$

where ι_α and ι_β are the embeddings of S^n in the mapping cones, one infers $\iota_\alpha \circ \beta = \iota_\alpha \circ \ell\alpha = \ell(\iota_\alpha \circ \alpha) = 0$, and therefore $C_{\alpha\beta} \simeq C_\alpha \vee S^m$. Similarly, $C_{\alpha\beta} \simeq C_\beta \vee S^m$, since $\alpha = \bar{\ell}\beta$ for some $\bar{\ell}$. Thus $C_\alpha \vee S^m \simeq C_\beta \vee S^m$. The resulting non-cancellation example of lowest dimension is as follows. Let $\omega \in \pi_6(S^3)$ be of order 12. Then

$$S^3 \cup_\omega e^7 \not\simeq S^3 \cup_{7\omega} e^7 \text{, but } (S^3 \cup_\omega e^7) \vee S^7 \simeq (S^3 \cup_{7\omega} e^7) \vee S^7$$

By passing to the dual situation, one obtains examples for non-cancellation of factors in a product. But these examples will not be finite dimensional, since the Eckmann-Hilton dual of a sphere S^n is an Eilenberg-MacLane space $K(\mathbb{Z},n)$, which is of infinite dimension for $n > 1$. It is therefore necessary to modify the dual situation. Hilton and Roitberg proceeded as follows. They considered principal G-bundles $G \to E_\alpha \to S^n$, classified by $\alpha \in \pi_{n-1}(G)$, G a connected Lie group. Supposing that α has order k and $\beta = \ell\alpha$ with ℓ prime to k, one wishes to infer

$$E_\alpha \times G \simeq E_\beta \times G \simeq E_{\alpha\beta},$$

where $E_{\alpha\beta}$ is defined by the pull-back diagram

$$\begin{array}{ccccc} E_{\alpha\beta} & \to & E_\beta & & \\ \downarrow & & \downarrow p_\beta & & \\ E_\alpha & \overset{p_\alpha}{\to} & S^n & \overset{\alpha^*}{\to} & BG \end{array}$$

The maps p_α and p_β are the natural projections, and α^* denotes the adjoint of $\alpha \in \pi_{n-1}(G)$. Of course, $\alpha^* \circ p_\alpha = 0$, and $\beta^* = \ell\alpha^*$. One would like to show that $\ell\alpha^* \circ p_\alpha = 0$, which would imply $E_{\alpha\beta} \simeq E_\alpha \times G$. The lack of a natural group structure in $[E_\alpha, BG]$ makes it difficult to compare $\ell\alpha^* \circ p_\alpha$ to $\alpha^* \circ p_\alpha$. Hilton and Roitberg found the following theorem to deal with the case $G = S^3$ (more general results may be found in the later papers [H150,H161,H186]).

<u>Theorem 4:</u> Let $\alpha \in \pi_{n-1}(S^3)$ and assume that $\ell\frac{(\ell-1)}{2}\omega \circ \Sigma^3 \alpha = 0$, where $\omega \in \pi_6(S^3)$ classifies $S^3 \to Sp(2) \to S^7$. Then $\ell\alpha^* \circ p_\alpha = 0$.

If we take $n = 7$ and $\alpha = \omega$, then $3(\omega \circ \Sigma^3\alpha) = 0$ since $\pi_9(S^3)$ has order 3. Thus $7\omega^* \circ p_\omega = 0$ and therefore $E_\omega \times S^3 \simeq E_{7\omega} \times S^3$; by definition $E_\omega = Sp(2)$. To see that $E_\omega \not\simeq E_{7\omega}$, one makes use of the homology decompositions of these spaces. By a result of I.M. James and J.H.C. Whitehead [11], $E_{n\omega} \simeq S^3 \cup_{n\omega} e^7 \cup e^{10}$. Thus $E_\omega \not\simeq E_{7\omega}$ since, as we observed earlier, $S^3 \cup_\omega e^7 \not\simeq S^3 \cup_{7\omega} e^7$.

The non-cancellation aspect of this example as well as the construction of new H-spaces become much more transparent, if one uses localization techniques. The localized fibrations

$$S_p^3 \to (E_\omega)_p \to S_p^7 , \quad \text{and}$$

$$S_p^3 \to (E_{7\omega})_p \to S_p^7$$

turn out to be fibre homotopy equivalent for each prime p ; the original fibrations belong thus to the same genus in the terminology of [H152]. In the more general setting of quasi-principal G-bundles

$$F \to E_i \to B , \quad i = 1,2$$

of the same genus, one can show that $E_1 \times F \simeq E_2 \times F$ (cf. [H161]; "quasi-principal" means that the bundle projection $p_i : E_i \to B$ composed with the classifying map $B \to BG$ of the associated principal G-bundle is 0-homotopic).

Recall that the genus set $G(X)$ of a nilpotent space X of finite type consists of all nilpotent homotopy types Y of finite type with p-localizations $X_p \simeq Y_p$ for all primes p . For instance $G(Sp(2)) = \{Sp(2), E_{7\omega}\}$. The genus sets of Lie groups provide a natural source for new H-spaces. Indeed one can show that if X is a finite H-complex, then every $Y \in G(X)$ is a finite H-complex (in [H186] it is proved that such a Y is a retract of $X \times X$; in the non-simply connected situation one needs to know that the Wall finiteness obstruction vanishes for Y , cf. [15]). A careful analysis of the construction of the members of a genus set reveals that if X is a loop space, then every $Y \in G(X)$ is a loop space [H186]. Therefore, all members of the genus $G(L)$ of a Lie group L have the homotopy type of topological groups. The genus sets of the simply connected rank two H-spaces are completely known. An interesting example is provided by the exceptional Lie group G_2 (cf. [H152]). The genus set of G_2 consists of precisely four members, $Y_1 = G_2$, Y_2, Y_3 and Y_4, whose cartesian powers are related by

$$Y_1^2 \simeq Y_2^2 \not\simeq Y_3^2 \simeq Y_4^2 , \quad Y_2^4 \simeq Y_3^4$$

The H-spaces of a fixed genus give always rise to non-cancellation examples of the type encountered in the original example of Hilton and Roitberg. One can show that if X_1, $X_2 \in G(X)$, X a finite H-complex, then there exists a product of spheres S, depending on X only, such that $X_1 \times S \simeq X_2 \times S$; conversely, if the finite H-complexes X_1 and X_2 stay in the relationship $X_1 \times S \simeq X_2 \times S$, with S a product of spheres, then one necessarily has $G(X_1) = G(X_2)$, cf. [14].

After the first examples of new H-spaces were known and understood, it was natural to study the homotopy classification of H-spaces with few cells. Peter Hilton was also involved in this project [H97,H112,H152]. His joint paper with Roitberg [H112] contains the following complete list of homologically torsion-free rank two H-spaces, modulo one ambiguity: $S^1 \times S^1$, $S^1 \times S^3$, $S^1 \times S^7$, $S^3 \times S^3$, $SU(3)$, $S^3 \times S^7$, $Sp(2)$, $(E_{2\omega})$, $E_{3\omega}$, $E_{4\omega}$, $E_{5\omega}$, $(E_{6\omega})$, $S^7 \times S^7$. The one doubt concerns $E_{2\omega}$ and $E_{6\omega}$, which were known to be both or neither H-spaces; they are not, as was proved by Zabrodsky [19] and, independently, by Sigrist and Suter using methods of K-theory [17].

At this point, it would seem natural to add a section on Peter Hilton's work in nilpotent spaces, which is intimately related to his current research in nilpotent groups. For instance, Hilton and Roitberg proved recently that nilpotent complexes of finite type are "Hopfian objects" in the homotopy category, generalizing the well known fact that finitely generated nilpotent groups are Hopfian. Influenced by a paper of J.M. Cohen [3], Hilton is studying jointly with Castellet and Roitberg "pseudo-identities", with applications to self maps of nilpotent spaces and homologically nilpotent fibrations. This work is still in progress and we will certainly hear more about it in the future.

REFERENCES

For the references labeled H see the complete list of Hilton's publications in this volume.

[1] J.F. Adams, Algebraic Topology - A Student's Guide, London Math. Soc. Lecture Notes Series 4, Cambridge University Press 1972.

[2] R. Bott and H. Samelson, On the Pontryagin product in spaces of paths, Comment. Math. Helv. 27 (1953), 320-337.

[3] J.M. Cohen, A spectral sequence automorphism theorem; applications to fibre spaces and stable homotopy, Topology 7 (1968), 173-177.

[4] T. Ganea, Cogroups and suspensions, Invent. Math. 9 (1970), 185-197.

[5] T. Ganea, Some problems on numerical homotopy invariants, Lecture Notes in Math. 249, Springer 1971, 13-22.

[6] M. Hall, A basis for free Lie rings and higher commutators in free groups, Proc. Amer. Math. Soc. 1 (1950), 575-581.

[7] P. Hall, A contribution to the theory of groups of prime power order, Proc. London Math. Soc. 36 (1934), 29-35.

[8] H. Hopf, Ueber Abbildungen von Sphären auf Sphären niedriger Dimension, Fund. Math. 25 (1935), 427-440.

[9] H. Hopf, Ueber die Topologie der Gruppen-Mannigfaltigkeiten und ihre Verallgemeinerung, Ann. of Math. (2) 42 (1941), 22-52.

[10] I.M. James, Reduced product spaces, Ann. of Math. 62 (1955), 170-197.

[11] I.M. James and J.H.C. Whitehead, The homotopy theory of sphere bundles over spheres: I, Proc. London Math. Soc. 4 (1954), 196-218.

[12] I. Kozma, Some relations between semigroups of polyhedra, Proc. Amer. Math. Soc. 39 (1973), 388-394.

[13] W. Magnus, Ueber Beziehungen zwischen höheren Kommutatoren, Journal für Math. (Crelle) 177 (1937), 105-115.

[14] G. Mislin, Cancellation properties of H-spaces, Comment. Math. Helv. 49 (2) 1974, 195-200.

[15] G. Mislin, Finitely dominated nilpotent spaces, Ann. of Math. 103 (1976), 547-556.

[16] H. Samelson, A connection between the Whitehead and the Pontryagin product, Amer. J. of Math. LXXV (1953), 744-752.

[17] F. Sigrist and U. Suter, Eine Anwendung der K-Theorie in der Theorie der H-Räume, Comment. Math. Helv. 47 (1) 1972, 36-52.

[18] J. Stasheff, Manifolds of the homotopy type of (non-Lie) groups, Bull. Amer. Math. Soc. 75 (1969), 998-1000.

[19] A. Zabrodsky, The classification of simply connected H-spaces with three cells, II: Math. Scand. 30 (1972), 211-222.

THE WORK OF PETER HILTON IN ALGEBRA

Urs Stammbach, ETH Zürich

Peter Hilton has published more than 250 articles and books; about 70 or 80 of them may be said to be of an algebraic nature. To give a description of the content of these papers in a few minutes is an impossible task. I shall therefore concentrate in this talk on a few aspects of Hilton's papers on algebra and try to describe the main ideas and the main results. I will not say anything about the numerous books and lecture notes Peter Hilton has written on various algebraic topics, although they undoubtedly constitute an equally important and influential part of his work. Many of the papers I shall be concerned with have one or sometimes two coauthors. Indeed, those who have had the chance to work with Peter Hilton know that he has an exceptional ability to stimulate and generate joint work.

For those who know Peter Hilton it is no surprise that most of his papers on algebra have a close relationship to algebraic topology; indeed either the motivation for the paper or the applications or even both lie in topology. Thus this part of his work can only be described in relation to topology and the assignment of a paper to algebra or topology is often arbitrary.

The first topic I want to talk about in some detail, the homotopy theory of modules, illustrates this point very clearly. The basic idea leading up to the theory was conceived in Zürich in 1955. Hilton, in his retrospective essay on duality in homotopy theory [H242], relates the story. In Warsaw Hilton had met Borsuk and had learned from him his definition of dependence of (continuous) maps between CW-complexes. The map $g : X \to Z$ is said to depend on $f : X \to Y$ if for every CW-complex $\bar{X}$ containing X such that f can be extended to $\bar{f} : \bar{X} \to Y$ the map g can be extended to a map $\bar{g} : \bar{X} \to Z$. Borsuk had proved that g depends on f if and only if there exists $h : Y \to Z$ such that g is homotopic to $h \circ f$. In the same year Hilton went to Zürich where on a suggestion of Beno Eckmann the two studied the same question for Λ-modules and in particular gave a sensible definition of homotopy of module maps. Since maps between modules may be subtracted, it is enough to define nullhomotopy. Borsuk's result suggested one define a homomorphism $\phi : A \to B$ to be nullhomotopic if it extends to every supermodule $\bar{A}$ of A. It is easy to see that this is equivalent to the requirement that ϕ factors through an injective supermodule $\bar{A}$ of A. The homotopy of module maps defined in this way was called injective or i-homotopy. The duality in module theory suggested then also a definition of projective homotopy: the map $\phi : A \to B$ is called p-nullhomotopic if it factors through a projective module P having B as quotient. It is interesting to note that Eckmann and Hilton have never published a joint paper on homotopy theory of modules; there are only two separate publications, namely transcripts of two talks, one given by Beno Eckmann in Louvain [3] and the other given by Peter Hilton in Mexico [H31]. In addition there is a short treatment of the theory in the book by Hilton on homotopy theory and duality [H77*]. These facts perhaps explain why the homotopy theory of modules has not become so widely known and employed in homological algebra as one might have expected. Let me briefly try to describe the main ideas.

If A,B are two Λ-modules, then the i-nullhomotopic maps $\phi : A \to B$ form a subgroup $\mathrm{Hom}_o(A,B)$ of $\mathrm{Hom}(A,B)$. As in topology Eckmann and Hilton define the i-homotopy group $\bar{\pi}(A,B)$ by

$$\bar{\pi}(A,B) = \mathrm{Hom}(A,B)/\mathrm{Hom}_o(A,B) \ .$$

If $\bar{A}$ is an injective module containing A (an "injective container" as Peter Hilton calls it in [H77*]) and $A \rightarrowtail \bar{A} \twoheadrightarrow \bar{A}/A$ the corresponding short exact sequence, then again as in topology $\bar{A}$ is called a cone CA of A and $\bar{A}/A$ a suspension ΣA of A. It is easy to see that the homotopy group $\bar{\pi}(\bar{A}/A,B)$ is independent of the chosen injective $\bar{A}$. This leads then to the n-th i-homotopy group by setting

$$\bar{\pi}_n(A,B) = \bar{\pi}(\Sigma^n A,B)$$

where $\Sigma^n A$ is the n-fold suspension of A. For $n \geq 2$ we have

$$\bar{\pi}_n(A,B) = H_{n-1}(\mathrm{Hom}(\underline{I},B))$$

where $\underline{I}$ is an injective resolution of A. One can then define relative i-homotopy groups and one obtains the usual exact sequences. In particular it is easy to see that any short exact sequence $A' \rightarrowtail A \twoheadrightarrow A''$ of modules gives rise to an associated long exact sequence

$$\to \bar{\pi}_{n+1}(A',B) \to \bar{\pi}_n(A'',B) \to \bar{\pi}_n(A,B) \to \bar{\pi}_n(A',B) \to \ldots$$

$$\ldots \to \bar{\pi}_1(A',B) \to \bar{\pi}_o(A'',B) \to \bar{\pi}_o(A,B) \to \bar{\pi}_o(A',B) \to \mathrm{Ext}^1(A'',B) \to \ldots$$

$$\ldots \to \mathrm{Ext}^{n-1}(A',B) \to \mathrm{Ext}^n(A'',B) \to \mathrm{Ext}^n(A,B) \to \mathrm{Ext}^n(A',B) \to \ldots$$

In [4] Eckmann and Kleisli have shown that in the case where Λ is a Frobenius algebra this sequence is essentially the complete homology sequence.

In 1960 a paper by Heller appeared [7] where analogous results for the homotopy theory in an arbitrary abelian category are obtained. (Since Heller works with p-homotopy he has to suppose that the category has enough projectives.) The notation ΩA for the kernel of a projective presentation

$P \twoheadrightarrow A$ of A was the same as that used by Eckmann and Hilton. In 1961 Heller published a further note [8] where he proved that in a category of modules over an Artin algebra, the fact that A is indecomposable implies that ΩA is indecomposable provided the projective presentation is minimal. This remark has become extremly important in representation theory, and $\Omega(-)$ is nowadays called the Heller operator.

In a certain sense the work on homotopy theory of modules was continued by Peter Hilton in a joint paper [H49] with Rees in 1961. Here the authors work in an abelian category $\underline{\underline{C}}$ with enough projectives and specialize to an appropriate category of modules only if necessary. This more general viewpoint had been made possible by developments between 1956 and 1960, in part by joint work of Hilton and Ledermann [H22], [H32], [H34]. They defined a categorical structure, called a homological ringoid, in which the basic notions of homological algebra could be defined. But let me return to the paper by Hilton and Rees. Their aim was to obtain an elementary homological proof of a result of Swan: A group G has periodic cohomology of period q if and only if $\mathbb{Z}$ has a G-projective resolution of period q. Of course, a group G is said to have periodic cohomology of period q if the functors $\mathrm{Ext}^i_G(\mathbb{Z},-)$ and $\mathrm{Ext}^{i+q}_G(\mathbb{Z},-)$ are naturally equivalent for all $i \geq 1$. This led the authors to consider quite generally the natural transformations between $\mathrm{Ext}^p(B,-)$, $B \in \underline{\underline{C}}$ and $F(-)$ where F is an additive functor from $\underline{\underline{C}}$ to the category of abelian groups. They showed that for $p > 0$

$$\mathrm{Nat}(\mathrm{Ext}^p(B,-)\ ,\ F(-)) = S_pF(-)\ ,$$

where $S_pF(-)$ denotes the p-th left satellite of F. This is a far reaching generalization of what is known as Yoneda's lemma (which appears as Lemma 1.1 in this paper). For $F(-) = A \otimes -$, this gives a characterisation of the Tor-functor

$$\mathrm{Tor}_p(A,B) = \mathrm{Nat}(\mathrm{Ext}^p(B,-),A \otimes -) ,$$

and for $F(-) = \mathrm{Ext}^1(A,-)$ it yields

$$\mathrm{Nat}(\mathrm{Ext}^1(B,-), \mathrm{Ext}^1(A,-)) = \underline{\pi}(A,B)$$

where $\underline{\pi}(A,B)$ is the p-homotopy group of Eckmann-Hilton.

In modern language one would describe this last result as follows. The assignment

$$A \rightsquigarrow \mathrm{Ext}^1(A,-)$$

defines a (contravariant) functor from the category $\underline{\underline{C}}$ to the category of additive functors from $\underline{\underline{C}}$ to abelian groups. The objects $\mathrm{Ext}^1(A,-)$ may be identified with the p-homotopy classes of objects in $\underline{\underline{C}}$ (in the sense of Eckmann-Hilton) or the classes of quasi-isomorphic objects (in the sense of Hilton-Rees). The morphism group between two such objects, i.e. $\mathrm{Nat}(\mathrm{Ext}^1(B,-) , \mathrm{Ext}^1(A,-))$ is $\underline{\pi}(A,B)$, the p-homotopy group of Eckmann-Hilton. I wanted to include this modern description of part of this paper, since in 1975 Auslander-Reiten [1] used this functor and the above result (as well as many other things) in their proof of the existence of almost split sequences. These sequences have proved in the mean time to be an indispensable tool in modern representation theory of Artin algebras.

Let me go back to homotopy theory. I have tried to explain how an attempt to copy certain ideas in topology into module theory led to the interesting notion of i-homotopy of modules. I have already also said that in module theory there is an obvious dual notion, that of p-homotopy. One of the key ideas of Eckmann-Hilton was to translate the dual notions in module theory back into topology. This led to the celebrated Eckmann-Hilton duality. But this really is topology, and no doubt a substantial part of Mislin's essay on Hilton's work in topology [10] will be devoted to a discussion of that duality. What is important here, is that this idea also gave rise to a development in category theory. I mean of course the series of papers, again jointly with Beno Eckmann on group-like structures [H51], [H56], [H57]. Starting from the wellknown fact that the set of homotopy classes of maps $\pi(X,Y)$ form a group if Y is a "group up to homotopy", the authors developed the notion of a group-like structure in a general category $\underline{\underline{C}}$.

A group-like structure on the object C in $\underline{\underline{C}}$ is a morphism $m : C \times C \to C$ which makes the set $\underline{\underline{C}}(X,C)$ into a group, naturally in X . In the three papers the two authors adopted an abstract categorical approach. The return they gained from this was enormous. First of all it made available to them the general duality principle of category theory, thus putting for example the notion of a cogroup in a category on the same level as the notion of a group. Secondly, it directed attention, in general and in applications primarily to those functors which respect certain constructions and which therefore carry the group-like structures from one category to another. This proved very fruitful from a heuristic point of view.

As an (easy) example of a result that may be found in these papers I mention the following. If X is a cogroup in $\underline{\underline{C}}$ and Y is a group in $\underline{\underline{C}}$, then $\underline{\underline{C}}(X,Y)$ inherits two natural group structures, one coming from X and the other coming from Y . It is shown in these papers that quite generally the two group structures agree, and that they are commutative. In particular, it follows that different cogroup structures on X and different group structures on Y give rise to one and the same abelian group $\underline{\underline{C}}(X,Y)$. This of course generalizes the fact that the fundamental group of a topological group G is abelian and that its group operation may be defined using the group operation of G .

In the course of these three papers the authors introduced and discussed many basic notions of category theory, like direct and inverse product, equalizer and coequalizer, intersection and union, adjoint functor etc. In reading these papers again I was reminded of my student days, when I learned a lot of category theory from them. We referred to these papers as the "trilogy by the pope and the copope". It is perhaps one of the nicest comments which can be made about a mathematical paper, that twenty years after its publication its content is common knowledge of working mathematicians. Such a comment certainly applies to these three papers.

The next algebraic topic Peter Hilton addressed was the theory of spectral sequences. Again jointly with Beno Eckmann he published a series of papers on the subject. Spectral sequences had been around for several years and had proved to be a very important tool in both topology and algebra. But a readable account of the theory that covered the many formally distinct cases was still missing.

The basic idea was distilled from their previous work on homotopy and duality. Basic to that theory is the notion of a composition functor T, i.e. a graded functor defined on a category of maps to the category of abelian groups and such that a factorization of a map f, $f = g \circ h$ gives rise to an exact sequence

$$\dots \to T_q(g) \to T_q(f) \to T_q(h) \to T_{q-1}(g) \to \dots$$

In [H68] the authors generalized this and considered infinite factorizations,

$$f = \dots \circ j_{p+1} \circ j_p \circ j_{p-1} \circ \dots .$$

Obvious examples are the skeleton decomposition of a simplicial complex, or the Postnikov decomposition of a 1-connected space. The authors showed that such a factorization, together with a composition functor gives rise to a spectral sequence. It is obtained via the associated exact couple and the Rees system. The treatment leads to results even in case the spectral sequence does not converge in the usual sense. In that case the E_∞-term of the spectral sequence is not isomorphic to the graded group associated with T(f) , but the theory allows to identify the deviation from an isomorphism.

A special feature of the authors approach to spectral sequences is that they work in an ungraded category so long as possible and only introduce the grading where it is really needed, for example for the discussion of convergence questions. This not only facilitates the presentation enormously, but also makes the whole theory more transparent. A student can now learn the subject without getting lost in a sea of indices.

In order to be able to subsume all the spectral sequences under one and the same theory they worked in an arbitrary abelian category [H70]. This had the further advantage that full use of the duality principle could be made. However in order to be able to proceed in this generality, the authors had to solve certain new problems of a more categorical nature; some of these have given rise to separate publications. In [H74] for example the notions of filtration, associated graded object and completion are discussed in great detail. Here direct and inverse limits are introduced in the framework of Kan's theory of adjoint functors. In [H86] they examine the curious fact that the E_∞-term of a spectral sequence is obtained as a limit followed by a colimit and also as a colimit followed by a limit. In general this procedure does not yield the same object; it does however in the special case that appears in the theory of spectral sequences. This question led to problems in category theory involving pushout and pullback squares. For these questions Peter Hilton was well prepared, for in 1965 at the conference on categorical algebra in La Jolla he had given a talk on "Correspondences and Exact Squares", a topic which is related to pullback (cartesian) and pushout (cocartesian) squares [H73]. In trying to describe the higher differentials of a spectral sequence Peter Hilton was led to consider correspondences (relations) between objects of an abelian category $\underline{\underline{A}}$. These define a new category $\tilde{\underline{\underline{A}}}$. By introducing the device of an exact square in $\underline{\underline{A}}$ Hilton was able to give a lucid description of the algebra of correspondences and of the way in which $\tilde{A}$ is obtained from $\underline{\underline{A}}$. He showed that there is a strong analogy with the classical procedure for passing from integers to fractions. Indeed, composition of morphisms in $\tilde{\underline{\underline{A}}}$ for example, corresponds to multiplication of fractions, etc.

I would like to add a further remark on this paper. It is the only research paper I know of which contains a reference to work done in elementary school. The reference is to a project of how to teach fractions to children. This then perhaps is a good place to say that it is impossible to fully appreciate the mathematician Peter Hilton without taking into account his concern with all aspects of education in mathematics, be it in elementary school or in graduate school. His publications and his lectures are vivid evidence of his conviction that teaching mathematics is an integral part of doing mathematics.

Again I break the chronological order to mention a further paper on spectral sequences. In [H165] of which I am a coauthor the naturality of the Lyndon-Hochschild-Serre spectral sequence for the homology of a group extension is exploited to yield results on the torsion of the differentials of the spectral sequence. Let $A \rightarrowtail G \twoheadrightarrow Q$ be a group extension with abelian kernel and let $\ell : A \to A$ be multiplication by the natural number ℓ , then the diagram

$$\begin{array}{rcccccc} \zeta : & A & \rightarrowtail & G & \twoheadrightarrow & Q \\ & \ell\downarrow & & \downarrow & & \| \\ \ell_*(\zeta) : & A & \rightarrowtail & \tilde{G} & \twoheadrightarrow & Q \end{array}$$

induces a map between the two associated spectral sequences. In particular, if $\ell_*(\zeta) = \zeta$ one obtains an automorphism of the spectral sequence. This can be used to deduce properties of its differentials. As an example I mention the case, where the extension splits. Then clearly $\zeta = 0 = \ell_*(\zeta)$. Hence the following square is commutative

$$\begin{array}{ccc} H_i(Q,H_1(A,\mathbb{Z})) & \xrightarrow{d_2} & H_{i-2}(Q,H_2(A,\mathbb{Z})) \\ \ell_* \downarrow & & \downarrow \ell_* \\ H_i(Q,H_1(A,\mathbb{Z})) & \xrightarrow{d_2} & H_{i-2}(Q,H_2(A,\mathbb{Z})) \end{array}$$

Choosing $\ell = 2$ and noting that 2_* on the left is multiplication by 2 and 2_* on right is multiplication by 4 one obtains $2d_2 = 0$. This result had been proved earlier by Evens [5] and Charlap-Vasquez [2] by rather complicated arguments. This technique and extensions of it produced various results on the torsion of the differentials of the Lyndon-Hochschild-Serre spectral sequence and eventually also estimates on the size of certain homology groups.

In another series of papers Peter Hilton was concerned with central group extensions and homology. If $N \rightarrowtail G \twoheadrightarrow Q$ is a central extension, then Ganea had shown using topological methods that the familiar five term sequence in integral homology can be extended by one term to the left:

$$(*) \qquad N \otimes G_{ab} \to H_2G \to H_2Q \to N \to G_{ab} \to Q_{ab} \to 0 .$$

In the first paper of the series [H140], jointly with Eckmann, the spectral sequence of the fibre sequence

$$K(G,1) \to K(Q,1) \to K(N,2)$$

was used to obtain a longer exact sequence in homology starting with H_4Q and containing Ganea's result as a corollary. A second line of attack is described in a paper by Eckmann, Hilton and myself [H136] where the map $\chi : N \otimes G_{ab} \to H_2G$ is identified with a commutator map in a free presentation of G. For extensions of a special kind (weak stem extensions) the sequence (*) was extended two further terms to the left

$$(**) \qquad H_3G \to H_3Q \to N \otimes G_{ab} \to H_2G \to H_2Q \to N \to G_{ab} \to Q_{ab} \to 0 .$$

This opened the way to an efficient and elegant approach to Schur's theory of covering groups, including Kervaire's results. In a second paper [H139] of the three authors a variant of the sequence (**) for arbitrary central extensions was obtained and in [H155] the sequence (*) was used to yield information about H_2Q, if G is a direct product, so that H_2G may be assumed to be known.

In [H162] Hilton and I worked on the problem of obtaining a sequence analogous to (*) without the hypothesis that N be central. Nomura [11] had already obtained such a sequence by topological methods. We proceeded algebraically by exploiting the naturality of the Lyndon-Hochschild-Serre spectral sequence and were able to generalize Nomura's result to arbitrary Q-coefficients. By a further analysis of the spectral sequence we also obtained an exact sequence for central extensions starting with H_4Q, different from the one described in [H140]. I would finally like to remark that these sequences for central extensions which I have mentioned above as well as their interrelationship have been carefully discussed in a paper by Gut [6].

A long series of articles of Hilton with various coauthors is devoted to the study of localizations of nilpotent groups and spaces. In the early seventies Bousfield-Kan and Sullivan defined a localization for topological spaces. The proper setting for this theory is the category of nilpotent spaces; a space X is called nilpotent if its fundamental group $\pi_1(X)$ is nilpotent and acts nilpotently on the higher homotopy groups $\pi_i(X)$, $i \geq 2$. If $X = K(G,1)$ with G a nilpotent group, the localization of X at a family P of primes is an aspherical space X_P. Hence $X_P = K(G_P,1)$ for some group G_P, the (algebraic) P-localization of G. If G is abelian, G_P is nothing else but the ordinary localization of G, known from commutative algebra. Various algebraic questions suggested themselves; I will only be able to mention a few. Firstly a connection between localization and homology surfaced, in that a nilpotent group G is P-local, i.e. isomorphic to its own P-localization if and only if its integral homology in positive dimensions is P-local. Starting from this Peter Hilton in [H167] described an elegant homological proof of the existence of the P-localization G_P of a nilpotent group G by an induction on the nilpotency class. In his approach he used as part of the inductive process the key fact that the natural map $\ell : G \to G_P$ is a P-equivalence, meaning that $\ker \ell$ is a P'-group and that there exists to every $x \in G_P$ a P'-number n such that $x^n \in \ell(G)$. In further papers, some of them joint, Peter Hilton addressed the following questions: Which group theoretic constructions are compatible with the localization map ℓ? What are the classes of groups in which a localization with good properties can be defined? What is the relation of localization to other group theoretic notions like the notion of the isolator of a subgroup?

Let me finally briefly describe the content of the two papers [H187], [H188], which are the product of joint work with Guido Mislin. Here some remarkable squares of nilpotent groups are discussed, namely squares of homomorphisms in the category $\underline{\underline{N}}$ of nilpotent groups

$$\begin{array}{ccc} G & \xrightarrow{\phi} & H \\ \psi \downarrow & & \downarrow \rho \\ K & \xrightarrow[\sigma]{} & L \end{array}$$

in which ϕ , σ are P-equivalences and ψ , ρ are Q-equivalences where P and Q are two families of primes with $P \cup Q = \pi$ the family of all primes. Such squares arise naturally in localization theory, for whenever one has a nilpotent group G the localizing maps form such a square

$$\begin{array}{ccc} G & \rightarrow & G_P \\ \downarrow & & \downarrow \\ G_Q & \rightarrow & G_{P\cup Q} \end{array} .$$

In [H187] Hilton and Mislin show that every such square is simultaneously a pullback and a pushout square in $\underline{\underline{N}}$. This is rather surprising since in $\underline{\underline{N}}$ pushout squares do not in general exist. This result was applied in [H188] to the study of the genus of a finitely generated nilpotent group with finite commutator subgroup. If N is such a group the genus G(N) is defined as the set of isomorphism classes of finitely generated nilpotent groups M with localizations $M_p \cong N_p$ for all primes p . Mislin [9] had previously shown that G(N) is finite. Here an abelian group structure in G(N) is defined. If H and K are two groups in G(N) then maps $\phi : N \rightarrow H$ and $\psi : N \rightarrow K$ and families P and Q of primes with $P \cup Q = \pi$ are constructed such that ϕ is a P-equivalence and ψ is a Q-equivalence. The addition in G(N) is then defined by constructing the pushout square

$$\begin{array}{ccc} N & \xrightarrow{\phi} & H \\ {\scriptstyle\psi}\downarrow & & \downarrow \\ K & \rightarrow & L \end{array} .$$

The group L is a representative of the sum in G(N) of the genus class of H and the genus class of K . The abelian group defined in this way can for example be used to estimate the size of G(N) .

Of course Peter Hilton's mathematical work continues. We certainly shall see in the future, as we have in the past, a lot of significant papers carrying that special trade mark: Peter Hilton.

I have tried in this talk to describe some of Peter Hilton's work in algebra. However it is not possible to give in such a small space a real impression of the richness of his work and of the immense influence it has had on the mathematical community. There is another shortcoming too. I have said nothing about Peter Hilton as a person, nothing about his gentleness and helpfulness. To give just a glimpse of that aspect of his personality I would like to close this talk with a description of my first personal encounter with Peter. It was in spring of 1965; I was at that time a rather shy graduate student at ETH who had just obtained some first results on his PhD thesis topic. Peter Hilton who was spending a couple of weeks in Zürich, showed interest in my work and asked me to describe some of the details. We arranged to meet in front of the Forschungsinstitut at Zehnderweg. At the appropriate time I was there, waiting very nervously for my meeting with an eminent and important mathematician. Then Peter arrived. His first question was: Which language shall we use: english, french or german? Naturally I was relieved to be able to opt for german. His second question, in german of course, was: Haben Sie Zeit, mit mir die London Times kaufen zu gehen? This is characteristic for Peter in more then one way. Not only is it evidence for the fact that Peter can hardly live without the daily issue of the London Times, but it also shows in a very subtle way Peter's thoughtfulness. No doubt, Peter had noticed my nervousness and thought that a walk and a little chit-chat would calm me down a bit. Indeed after our walk we had an intensive and satisfying talk on various mathematical matters.

REFERENCES

For the references labeled H see the complete list of Hilton's publications in this volume.

[1] M. Auslander and I. Reiten, Representation theory of Artin algebras III. Almost split sequences, Comm. Algebra 3 (1975), 239-294.

[2] L.S. Charlap and A.T. Vasquez, Characteristic classes for modules over groups I, Trans. Amer. Math. Soc. 137 (1969), 533-549.

[3] B. Eckmann: Homotopie et dualité, Colloque de topologie algébrique, Louvain 1956, 41-53.

[4] B. Eckmann and H. Kleisli, Algebraic homotopy groups and Frobenius algebras, Ill. J. Math. 6 (1962), 533-552.

[5] L. Evens, The Schur multiplier of a semi-direct product, Ill. J. Math. 16 (1972), 166-181.

[6] A. Gut, A ten-term exact sequence in the homology of a group extension, J. Pure Appl. Algebra 8 (1976).

[7] A. Heller, The loop space functor in homological algebra, Trans. Amer. Math. Soc. 96 (1960), 382-394.

[8] A. Heller, Indecomposable representations and the loop-space operation, Proc. Amer. Math. Soc. 12 (1961), 640-643.

[9] G. Mislin, Nilpotent groups with finite commutator subgroups. In: Localization in Group theory and Homotopy theory, Lecture Notes in Math. Vol. 418, Springer 1974, 103-120.

[10] G. Mislin, Essay on Hilton's work in topology, this volume.

[11] Y. Nomura, The Whitney join and its dual, Osaka J. Math. 7 (1970), 353-373.

THE DUAL WHITEHEAD THEOREMS

J. P. May
The University of Chicago

For Peter Hilton on his 60th birthday

Eckmann-Hilton duality has been around for quite some time and is something we now all take for granted. Nevertheless, it is a guiding principle to "the homotopical foundations of algebraic topology" that is still seldom exploited as thoroughly as it ought to be. In 1971, I noticed that the two theorems commonly referred to as Whitehead's theorem are in fact best viewed as dual to one another. I've never published the details. (They were to appear in a book whose title is in quotes above and which I contracted to deliver to the publishers in 1974; 1984, perhaps?) This seems a splendid occasion to advertise the ideas. The reader is referred to Hilton's own paper [2] for a historical survey and bibliography of Eckmann-Hilton duality. We shall take up where he left off.

The theorems in question read as follows.

Theorem A. A weak homotopy equivalence $e:Y \to Z$ between CW complexes is a homotopy equivalence.

Theorem B. An integral homology isomorphism $e:Y \to Z$ between simple spaces is a weak homotopy equivalence.

In both, we may as well assume that Y and Z are based and (path) connected and that e is a based map. The hypothesis of Theorem A (and conclusion of Theorem B) asserts that $e_*:\pi_*(Y) \to \pi_*(Z)$ is an isomorphism. The hypothesis of Theorem B asserts that $e_*:H_*(X) \to H_*(Y)$ is an isomorphism. A simple space is one whose fundamental group is Abelian and acts trivially on the higher homotopy groups. Theorem B remains true for nilpotent spaces, for which the fundamental group is nilpotent and acts nilpotently on the higher homotopy groups. More general versions have also been proven.

It is well understood that Theorem A is elementary. However, the currently fashionable proof of Theorem B and its generalizations depends on use of the Serre spectral sequence. We shall obtain a considerable generalization of Theorem B by a strict word for word dualization of the simplest possible proof of Theorem A, and our arguments will also yield a generalized form of Theorem A.

We shall work in the good category $\mathcal{J}$ of compactly generated weak Hausdorff based spaces. Essentially the same arguments can be carried out in other good topological categories, for example, in good categories of G-spaces, or spectra, or G-spectra. An axiomatic setting could be developed but would probably obscure the simplicity of the ideas.

We shall use very little beyond fibre and cofibre sequences. Let $X \wedge Y$ be the smash product $X \times Y/X \vee Y$ and let $F(X,Y)$ be the function space of based maps $X \to Y$. The source of duality is the adjunction homeomorphism

(1) $$F(X \wedge Y,Z) \cong F(X,F(Y,Z)).$$

Let $CX = X \wedge I$, $\Sigma X = X \wedge S^1$, $PX = F(I,X)$, and $\Omega X = F(S^1,X)$, where I has basepoint 1 in forming CX and 0 in forming PX. For a based map $f:X \to Y$, let $Cf = Y \cup_f CX$ be the cofibre of f and let $Ff = X \times_f PY$ be the fibre of f. Let $\pi(X,Y)$ denote the pointed set of homotopy classes of based maps $X \to Y$. For spaces J and K, we have the long exact sequences of pointed sets (and further structure as usual)

(2) $$\cdots \longrightarrow \pi(\Sigma^n Cf,K) \longrightarrow \pi(\Sigma^n Y,K) \longrightarrow \pi(\Sigma^n X,K) \longrightarrow \pi(\Sigma^{n-1}Cf,K) \longrightarrow \cdots ;$$

(3) $$\cdots \longrightarrow \pi(J,\Omega^n Ff) \longrightarrow \pi(J,\Omega^n X) \longrightarrow \pi(J,\Omega^n Y) \longrightarrow \pi(J,\Omega^{n-1}Ff) \longrightarrow \cdots .$$

The crux of Theorem A is the following triviality; we shall give the proof since nothing else requires any work.

Lemma 1. Let $e:Y \to Z$ be a map such that $\pi(J,Fe) = 0$. If $hi_1 = eg$ and $hi_0 = fi$ in the following diagram, where i_0, i_1, and i are the evident inclusions, then there exist $\tilde{g}$ and $\tilde{h}$ which make the diagram commute.

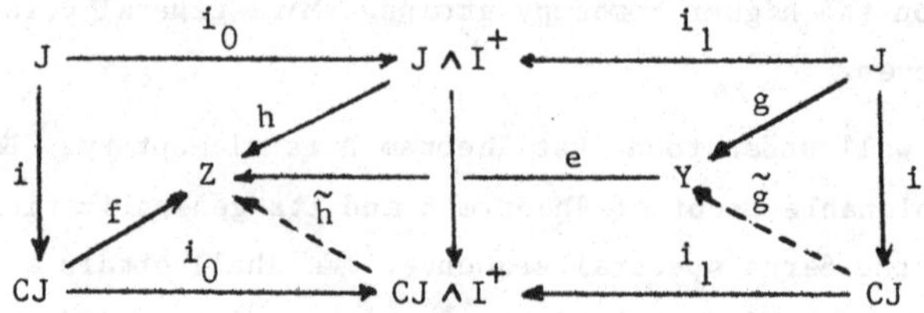

Proof. Define $k_0:J \to Fe$ by $k_0(j) = (g(j), \omega_0(j))$, where $\omega_0(J) \in PZ$ is specified by

$$\omega_0(j)(s) = \begin{cases} f(j,\ 1-2s) & \text{if } s \leq 1/2 \\ h(j,\ 2s-1) & \text{if } s \geq 1/2 \ . \end{cases}$$

Choose a homotopy $k:J \wedge I^+ \to Fe$ from k_0 to the trivial map and define $\tilde{g}:CJ \to Y$ and $\omega:J \wedge I^+ \to PZ$ by

$$k(j,t) = (\tilde{g}(j,t), \omega(j,t)).$$

Define $\tilde{h}:CJ \wedge I^+ \to Z$ by

$$\tilde{h}(j,s,t) = \omega(j,u(s,t))(v(s,t)),$$

where $u(s,t) = \min(s,2t)$ and $v(s,t) = \max(\frac{1}{2}(1+t)(1-s), 2t-1)$. Then $\tilde{g}$ and $\tilde{h}$ make the diagram commute.

We now introduce a general version of cellular theory.

Definition 2. Let $\mathcal{J}$ be any collection of spaces such that $\Sigma J \in \mathcal{J}$ if $J \in \mathcal{J}$. A map $e:Y \to Z$ is said to be a weak $\mathcal{J}$-equivalence if $e_*:\pi(J,Y) \to \pi(J,Z)$ is a bijection for all $J \in \mathcal{J}$. A $\mathcal{J}$-complex is a space X together with subspaces X_n and maps $j_n:J_n \to X_n$, $n \geq 0$, such that $X_0 = \{*\}$, J_n is a wedge of spaces in $\mathcal{J}$, $X_{n+1} = Cj_n$, and X is the union of the X_n. The evident map from the cone on a wedge summand of J_{n-1} into X is called an n-cell. The restriction of j_n to a wedge summand is called

is a product of spaces in $\mathcal{K}$, $X_{n+1} = Fk_n$, and X is the inverse limit of the X_n (via the given maps). The evident map from X to the paths on a factor of K_{n-1} is called an n-cocell. The projection of k_n to a factor is called a coattaching map. A map $p: X \to A$ is said to be a projection onto a quotient tower if A is a $\mathcal{K}$-tower, p is the inverse limit of maps $X_n \to A_n$, and the composite of p and each n-cocell $A \to A_n \to PK$ is an n-cocell of X.

Example 3*. Let $\mathcal{A}$ be any collection of Abelian groups which contains $\{0\}$, for example the collection $\mathcal{Ab}$ of all Abelian groups. Let $\mathcal{KA}$ be the collection of all Eilenberg-MacLane spaces K(A,n) such that $A \in \mathcal{A}$ and $n \geq 0$. (We require Eilenberg-MacLane spaces to have the homotopy types of CW-complexes; this doesn't effect closure under loops by a theorem of Milnor.) A $\mathcal{KAb}$-tower X such that K_n is a $K(\pi_{n+1}, n+2)$ for $n \geq 0$ is called a simple Postnikov tower and satisfies $\pi_n(X) = \pi_n$. Its coattaching map k_n is usually written k^{n+2} and called a k-invariant.

Theorem 4* (coHELP). Let A be a quotient tower of a $\mathcal{K}$-tower X and let $e: Y \to Z$ be a weak $\mathcal{K}$-equivalence. If $p_1 h = ge$ and $p_0 h = pf$ in the following diagram, then there exist $\tilde{g}$ and $\tilde{h}$ which make the diagram commute.

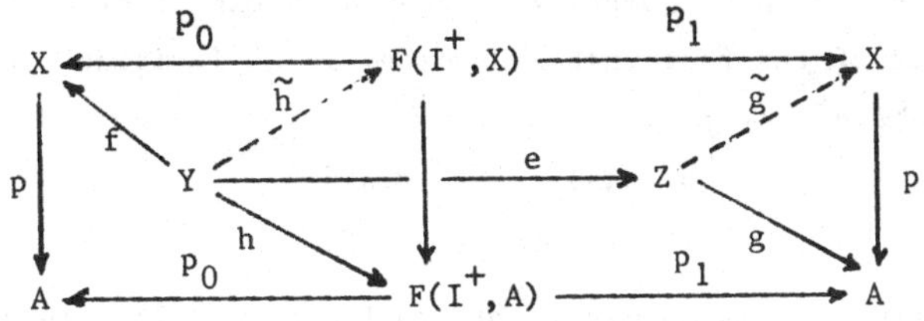

Proof. By (1) and (2) and the fact that $\mathcal{K}$ is closed under loops, the hypothesis implies that $\pi(Ce, K) = 0$ for all $K \in \mathcal{K}$. The conclusion follows inductively by a cocell by cocell application of Lemma 1*.

In particular, the projection $p: X \to A$ is a fibration.

Theorem 5*. For every weak $\mathcal{K}$-equivalence $e: Y \to Z$ and every $\mathcal{K}$-tower X, $e^*: \pi(Z, X) \to \pi(Y, X)$ is a bijection.

Proof. The surjectivity and injectivity of e^* result by application of coHELP to the quotient towers $X \to *$ and $F(I^+, X) \to F(\partial I^+, X)$, respectively.

The cocellular Whitehead theorem is a formal consequence.

Theorem 6* (Whitehead). Every weak $\mathcal{K}$-equivalence between $\mathcal{K}$-towers is a homotopy equivalence.

To derive useful conclusions from these theorems we have to use approximations of spaces by CW-complexes and by Postnikov towers. For a space X of the homotopy type of a CW-complex, we have

$$\tilde{H}^n(X;A) = \pi(X,K(A,n)).$$

However, $\mathcal{K}$-towers hardly ever have the homotopy types of CW-complexes. The best conceptual way around this is to pass from the homotopy category $h\mathcal{J}$ to the category $\bar{h}\mathcal{J}$ obtained from it by inverting its weak homotopy equivalences. For any space X, there is a CW-complex ΓX and a weak homotopy equivalence $\gamma:\Gamma X \to X$. The morphisms of $\bar{h}\mathcal{J}$ can be specified by

$$[X,Y] = \pi(\Gamma X, \Gamma Y),$$

with the evident composition. By Theorem 5, we have $[X,Y] = \pi(X,Y)$ if X has the homotopy type of a CW-complex. Either as a matter of definition or as a consequence of the fact that cohomology is an invariant of weak homotopy type, we have

$$\tilde{H}^n(X;A) = [X,K(A,n)]$$

for any space X.

Now return to Example 3*. Say that a map $e:Y \to Z$ is an $\mathcal{A}$-cohomology isomorphism if $e^*:H^*(Z;A) \to H^*(Y;A)$ is an isomorphism for all $A \in \mathcal{A}$. If Y and Z are CW-complexes, then e is an $\mathcal{A}$-cohomology isomorphism if and only if it is a weak $\mathcal{K}\mathcal{A}$-equivalence.

Theorem 5#. For every $\mathcal{A}$-cohomology isomorphism $e:Y \to Z$ and every $\mathcal{K}\mathcal{A}$-tower X, $e^*:[Z,X] \to [Y,X]$ is a bijection.

Proof. We may as well assume that Y and Z are CW-complexes, and the result is then a special case of Theorem 5*.

This leads to the cohomological Whitehead Theorem.

Theorem $6^{\#}$ (Whitehead). The following statements are equivalent for a map $e:Y \to Z$ in $\overline{h\mathcal{J}}$ between connected spaces Y and Z of the weak homotopy type of $\mathcal{K}\mathcal{A}$-towers.

(1) e is an isomorphism in $\overline{h\mathcal{J}}$.

(2) $e_*:\pi_*(Y) \to \pi_*(Z)$ is an isomorphism.

(3) $e^*:H^*(Z;A) \to H^*(Y;A)$ is an isomorphism for all $A \in \mathcal{A}$.

(4) $e^*:[Z,X] \to [Y,X]$ is a bijection for all $\mathcal{K}\mathcal{A}$-towers X.

If $\mathcal{A}$ is the collection of modules over a commutative ring R, then the following statement can be added to the list.

(5) $e_*:H_*(Y;R) \to H_*(Z;R)$ is an isomorphism.

Proof. The previous theorem gives (3) $\Longrightarrow$ (4), (4) $\Longrightarrow$ (1) is formal, and (1) $\Longleftrightarrow$ (2) by the definition of $\overline{h\mathcal{J}}$; (2) $\Longrightarrow$ (3) and (2) $\Longrightarrow$ (5) since homology and cohomology are invariants of weak homotopy type, and (5) $\Longrightarrow$ (3) by the universal coefficients spectral sequence.

When $\mathcal{A} = \mathcal{A}b$, the implication (5) $\Longrightarrow$ (2) is the promised generalization of Theorem B. It is almost too general. Given a space X, it is hard to tell whether or not X has the weak homotopy type of a $\mathcal{K}\mathcal{A}b$-tower. If X is simple, or nilpotent, the standard theory of Postnikov towers shows that X does admit such an approximation. However, in Definition 2^* with $\mathcal{K} = \mathcal{K}\mathcal{A}b$, each K_n can be an arbitrary infinite product of K(A,q)'s for varying q and the maps $k_n:X_n \to K_n$ are completely unrestricted. Thus $\mathcal{K}\mathcal{A}b$-towers are a great deal more general than nilpotent Postnikov towers; compare Dror [1].

The applicability of Theorem 6^* to general collections $\mathcal{A}$ is of considerable practical value. A space X is said to be $\mathcal{A}$-complete if $e^*:[Z,X] \to [Y,X]$ is a bijection for all $\mathcal{A}$-cohomology isomorphisms $e:Y \to Z$. Thus Theorem $5^{\#}$ asserts that $\mathcal{K}\mathcal{A}$-towers are $\mathcal{A}$-complete. The completion of a space X at $\mathcal{A}$ is an $\mathcal{A}$-cohomology isomorphism from X to an $\mathcal{A}$-complete space. For a set of primes T, the completion of X at the collection of T-local or T-complete Abelian groups is the localization or completion of X at T; in the latter case, we may equally well use the collection of those Abelian groups which are vector spaces over Z/pZ for some prime $p \in T$.

These ideas give the starting point for an elementary homotopical account of the theory of localizations and completions in which the latter presents little more difficulty than the former; compare Hilton, Mislin, and Roitberg [3]. Details should appear eventually in "The homotopical foundations of algebraic topology". The equivariant generalization of the basic constructions and characterizations has already been published [4,5], and there the present focus on cohomology rather than homology plays a mathematically essential role.

Bibliography

1. E. Dror. A generalization of the Whitehead theorem. Lecture Notes in Mathematics Vol. 249. Springer (1971), 13-22.
2. P. Hilton. Duality in homotopy theory: a rectrospective essay. J. Pure and Applied Algebra 19(1980), 159-169.
3. P. Hilton, G. Mislin, and J. Roitberg. Localization of nilpotent groups and spaces. North-Holland Math. Studies Vol. 15. North Holland. 1975.
4. J. P. May, J. McClure, and G. Triantafillou. Equivariant localization. Bull. London Math. Soc. 14(1982), 223-230.
5. J. P. May. Equivariant completion. Bull. London Math. Soc. 14(1982), 231-237.

HOMOTOPY COCOMPLETE CLASSES OF SPACES AND THE REALIZATION OF THE SINGULAR COMPLEX

D.Puppe
Mathematisches Institut
der Universität Heidelberg
D-6900 Heidelberg
Federal Republic of Germany

Dedicated to PETER JOHN HILTON on the occasion of his 60th birthday

We give a new proof for the well known theorem that the canonical map

$$\varepsilon_X\colon |SX| \longrightarrow X$$

from the geometric realization of the singular complex of a space X is a homotopy equivalence if X has the homotopy type of a CW-complex. The theorem is due to Giever [G, 6.Theorem VI] and Milnor [Mi, 4.Theorem 4]. Proofs of this or closely related theorems may also be found in [G-Z, VII], [L, VII.10.10] and [May, 16.6].

Our proof is elementary in the sense that we use only some basic geometric constructions (like the double mapping cylinder and barycentric subdivision) and e.g. do not need homotopy groups or homology.

For this purpose we study classes of (topological) spaces which are *homotopy cocomplete*, i.e. closed under all homotopy colimits, and we show in particular:

(A) *The class* E *of spaces* X *for which* ε_X *is a homotopy equivalence is homotopy cocomplete.*

(B) *The class* $\mathcal{W}$ *of spaces having the homotopy type of a CW-complex is contained in any homotopy cocomplete class which contains a one point space.*

Obviously (A) and (B) imply the Giever-Milnor theorem on ε_X and also the fact that $\mathcal{W}$ is the smallest homotopy cocomplete class which contains a one point space. This latter fact can, of course, also be seen directly.

Some simple categorical arguments give the corollary that ε_X is a weak homotopy equivalence for any space X (Section 3).

The author wishes to express his thanks to the Mathematical Institute of the National University of Mexico, where he was a guest when this work originated.

1. Homotopy cocomplete classes of spaces

We can work either in the category *Top* of all topological spaces or in some other category suitable for homotopy theory like compactly generated spaces. What one really needs about the underlying category is very little and can easily be abstracted from the following. For simplicity we talk about *Top* .

1.1. DEFINITION. Let C be a class of objects of *Top* (which we sometimes identify with the full subcategory of *Top* having those objects). C is called *homotopy cocomplete* if the following three conditions are satisfied:

(a) *If* $X \in C$ *and* X' *has the same homotopy type as* X *then* $X' \in C$.

(b) *Any topological sum (coproduct) of spaces in* C *is an element of* C . *(In particular* $\emptyset \in C$.*)*

(c) *The double mapping cylinder* $Z(f_1, f_2)$ *of any diagram*

$$X_1 \xleftarrow{f_1} X_o \xrightarrow{f_2} X_2$$

in C *is an element of* C .

Recall that $Z(f_1, f_2)$ is obtained from the sum

$$X_1 \sqcup (X_o \times I) \sqcup X_2$$

by the identifications

$$(x, 0) \sim f_1 x$$

$$(x, 1) \sim f_2 x$$

for all $x \in X_o$.

There is a general notion of homotopy colimit for any (small) diagram in *Top* [V,(1.1)]. It is not hard to show that C satisfies (a) - (c) if and only if it is closed under all such homotopy colimits; but we shall not use this in the present paper.

1.2. EXAMPLES. The following classes of spaces are homotopy cocomplete:

(i) The class $\{\emptyset\}$.

(ii) The class $\mathcal{W}$ of spaces having the homotopy type of a CW-complex.

(iii) The class $\mathcal{N}$ of spaces which are numerably locally contractible.

This is trivial for (i). For (iii) the necessary definitions and proofs may be found in [P, Section 1] where it is also observed that the class $\mathcal{N}$ is strictly larger than $\mathcal{W}$. About (ii) the following three things may be said:

1. The assertion is well known.
2. To check it one needs only to observe that conditions (a) and (b) are trivial for $\mathcal{W}$, whereas (c) follows from the "homotopy invariance" of the double mapping cylinder by "cellular approximation".
3. It will also be an immediate corollary of 1.3 and 2.1 below.

1.3. PROPOSITION. *Let $\mathcal{C}$ be a class of spaces which is homotopy cocomplete and contains a one point space. Then $\mathcal{W} \subset \mathcal{C}$, i.e. $\mathcal{C}$ contains all spaces of the homotopy type of a CW-complex.*

Proof. Observe first that each one point space pt belongs to $\mathcal{C}$ (by (a)) and hence each discrete space (by (b)). In particular $S^0 \in \mathcal{C}$. Using (c) we get $S^n \in \mathcal{C}$ for all n by induction because S^n is the double mapping cylinder of

$$pt \longleftarrow S^{n-1} \longrightarrow pt \ .$$

Now let A be any CW-complex and denote by A^n its n-dimensional skeleton. Then $A^0 \in \mathcal{C}$ because it is a discrete space. For any $n \geq 1$ one obtains A^n from A^{n-1} by attaching n-cells which can be described by saying that A^n is the double mapping cylinder of

$$A^{n-1} \xleftarrow{\ f\ } S^{n-1} \times D \xrightarrow{\ pr_2\ } D \ ,$$

where D is a suitable discrete space and f some

continuous map. Using (c) we get $A^n \in C$ for all n by induction, because we know already that D and $S^{n-1} \times D$ (which is the sum of copies of S^{n-1}) are in C.

Now A is the colimit of the sequence

$$A^o \subset A^1 \subset A^2 \subset \ldots$$

and because each inclusion map is a cofibration the colimit A is homotopy equivalent to the telescope (= homotopy colimit) $\tilde{A}$ of the same sequence [D1, 2.Lemma 6 and Remark 1]. But $\tilde{A}$ may be described as the double mapping cylinder of

$$\coprod_{n \text{ odd}} A^n \xleftarrow{f_1} \coprod_n A^n \xrightarrow{f_2} \coprod_{n \text{ even}} A^n$$

where f_1 maps the summand A^n into
A^n by the identity if n is odd
A^{n+1} by inclusion if n is even

and f_2 does the same with "odd" and "even" interchanged. By (b) all the spaces in the above diagram are in C. Hence $\tilde{A} \in C$ by (c) and $A \in C$ by (a). Since A was an arbitrary CW-complex we have $\mathcal{W} \subset C$ (using (a) once more).

2. The realization of the singular complex

Let X be a space. By SX we denote the "singular complex" of X considered as a simplicial set, i.e. SX consists of the sets of singular simplices

$$S_nX = \{\sigma: \Delta^n \longrightarrow X\}$$

together with face and degeneracy operators.

If K is any simplical set then $|K|$ denotes its geometric realization in the sense of Milnor [Mi]. It is obtained from

$$\coprod_n (K_n \times \Delta^n)$$

by the well known identifications using both faces and degeneracies.

For each space X one has a canonical map

$$\varepsilon_X\colon |SX| \longrightarrow X$$

which is induced by sending

$$(\sigma,t) \in S_nX \times \Delta^n$$

into $\sigma(t) \in X$.

2.1. THEOREM. *The class* $\mathcal{E}$ *of spaces* X *for which* ε_X *is a homotopy equivalence is homotopy cocomplete.*

If X is a one point space so is $|SX|$. Hence $\mathcal{E}$ contains all one point spaces. Combining 1.3 with 2.1 we get $\mathcal{W} \subset \mathcal{E}$. On the other hand $|SX|$ is always a CW-complex, which obviously implies $\mathcal{E} \subset \mathcal{W}$. Hence we have $\mathcal{W} = \mathcal{E}$ and we may state

2.2. COROLLARY. *The following three classes of spaces are the same:*

(1) *The class of spaces having the homotopy type of a CW-complex.*

(2) *The class of all* X *for which* $\varepsilon_X\colon |SX| \longrightarrow X$ *is a homotopy equivalence.*

(3) *The smallest class of spaces which is homotopy cocomplete and contains a one point space.*

In the rest of this section we give the proof of Theorem 2.1. We have to verify conditions (a), (b) and (c) of Definition 1.1 for the class $\mathcal{C} = \mathcal{E}$.

Condition (b) is trivially satisfied because the functors singular complex $X \longmapsto SX$ and geometric realization $K \longmapsto |K|$ both commute with arbitrary coproducts.

Condition (a) is not much harder. It follows from the fact that both functors preserve homotopies. To say this in more detail let $\Delta[n]$ be the standard simplicial set with

$$\Delta[n]_q = \{(\text{weakly) increasing maps } [q] \longrightarrow [n]\}$$

where $[n] = \{0,1,\ldots,n\}$. Then $|\Delta[n]| = \Delta^n$.

In fact we need only the case n = 1.

For any simplicial set K one has the canonical map (of simplicial sets)

$$\eta_K : K \longrightarrow S|K|$$

sending $x \in K_n$ into the singular simplex $\sigma\colon \Delta^n \longrightarrow |K|$ which maps $t \in \Delta^n$ into the equivalence class of (x,t) . In particular this gives

$$\Delta[1] \xrightarrow{\eta} S\Delta^1$$

which we use to form

$$\begin{aligned} |SX| \times \Delta^1 &= |SX| \times |\Delta[1]| \cong |SX \times \Delta[1]| \\ &\xrightarrow{|id\times\eta|} |SX \times S\Delta^1| \cong |S(X\times\Delta^1)| \quad . \end{aligned}$$

Let φ denote this composition. Now if

$$h\colon X \times \Delta^1 \longrightarrow Y$$

is a homotopy connecting $f,g : X \longrightarrow Y$ to each other, then

$$|Sh| \circ \varphi : |SX| \times \Delta^1 \longrightarrow |S(X\times\Delta^1)| \longrightarrow |SY|$$

is a homotopy connecting $|Sf|$ to $|Sg|$.

This shows that the composite functor $X \longmapsto |SX|$ preserves homotopies. (We need not talk explicitly about simplicial homotopies.) It follows that if $f\colon X \longrightarrow X'$ is a homotopy equivalence so is $|Sf|$, and the diagram

$$\begin{array}{ccc} |SX| & \xrightarrow{|Sf|} & |SX'| \\ {\scriptstyle \varepsilon_X}\downarrow & & \downarrow{\scriptstyle \varepsilon_{X'}} \\ X & \xrightarrow{f} & X' \end{array}$$

shows that condition (a) holds.

We now turn to condition (c) which is,

of course, the crux of the matter. Let

$$X_1 \xleftarrow{f_1} X_o \xrightarrow{f_2} X_2$$

be given. The double mapping cylinder $Z(f_1,f_2)$ which we abbreviate by Y is a quotient of

$$X_1 \sqcup (X_o \times I) \sqcup X_2 .$$

Let Y_1 be the image of

$$X_1 \sqcup (X_o \times [0,\tfrac{3}{4}])$$

in Y, and Y_2 the image of

$$(X_o \times [\tfrac{1}{4},1]) \sqcup X_2 .$$

Finally let $Y_o = Y_1 \cap Y_2$. Then $Y_o \cong X_o \times [\frac{1}{4},\frac{3}{4}]$ and there are obvious canonical homotopy equivalences

$$Y_i \xrightarrow{\simeq} X_i \quad , \quad i = 0,1,2 .$$

Now let us make the hypothesis that $X_i \in E$ for i = 0,1,2. This implies $Y_i \in E$, because we know already that E is closed under homotopy equivalences (condition (a)). We have to show that $Y \in E$, i.e. that

$$\varepsilon_Y\colon |SY| \longrightarrow Y$$

is a homotopy equivalence. Instead of doing this directly we consider a different kind of geometric realization. If K is a simplicial set we denote by $\|K\|$ the space obtained from

$$\coprod_n (K_n \times \Delta^n)$$

by making only those identifications which correspond to face operators, i.e. for each <u>strictly</u> increasing map $\alpha\colon [m] \longrightarrow [n]$ and all $x \in K_n$, $t \in \Delta^m$ we identify (α^*x,t) with $(x,\alpha t)$. (In the case $K = SX$ this goes back to Giever [G].) One has a canonical map

$$\|K\| \longrightarrow |K|$$

which is known to be a homotopy equivalence [D2, Proposition 1]. Composition with ε_X gives a canonical map

$$\varepsilon'_X : \|SX\| \longrightarrow X$$

for each space X , and it makes no difference whether we say that ε_X or that ε'_X is a homotopy equivalence.

We are going to show that ε'_Y is a homotopy equivalence. For this we consider the diagram

$$\begin{array}{ccccc} \|SY_1\| & \xleftarrow{\|Si_1\|} & \|SY_o\| & \xrightarrow{\|Si_2\|} & \|SY_2\| \\ \downarrow{\scriptstyle \varepsilon'_{Y_1}} & & \downarrow{\scriptstyle \varepsilon'_{Y_o}} & & \downarrow{\scriptstyle \varepsilon'_{Y_2}} \\ Y_1 & \xleftarrow{i_1} & Y_o & \xrightarrow{i_2} & Y_2 \end{array}$$

The maps i_1, i_2 are inclusions. All horizontal maps are cofibrations, and all vertical maps are homotopy equivalences. By the glueing lemma [B, 7.5.7] or [D1, 2.Lemma 1] one obtains a homotopy equivalence ε' between the pushouts of the two lines. The pushout of the lower line is obviously Y . The pushout of the upper line is the realization $\|SY_1 \cup SY_2\|$ of the sub-(simplicial set) $SY_1 \cup SY_2$ of SY consisting of those singular simplices of Y which lie entirely in Y_1 or entirely in Y_2 . We have a commutative diagram

$$\begin{array}{ccc} \|SY_1 \cup SY_2\| & \xrightarrow{\subset} & \|SY\| \\ & {\scriptstyle \varepsilon'} \searrow^{\simeq} \quad \swarrow {\scriptstyle \varepsilon'_Y} & \\ & Y & \end{array}$$

Thus in order to prove that ε'_Y is a homotopy equivalence it suffices to show that $\|SY_1 \cup SY_2\|$ is a deformation retract of $\|SY\|$. This is a special case of

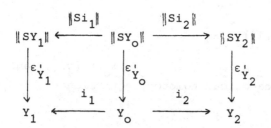

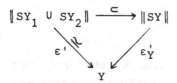

2.3. LEMMA. *Let* Y *be a space and* $\mathcal{U}$ *a set of subsets of* Y *such that the interiors of the elements of* $\mathcal{U}$ *cover* Y . *Let*

$$S(Y;\mathcal{U}) = \bigcup_{U\in\mathcal{U}} SU \subset SY .$$

Then $\|S(Y;\mathcal{U})\|$ *is a strong deformation retract of* $\|SY\|$.

Proof. In the standard setup of singular homology one proves the excision theorem by showing that

$$S(Y;\mathcal{U}) \xrightarrow{\subset} SY$$

induces a homology isomorphism. What we do here is just translate this proof (whose main tool is barycentric subdivision) into homotopy.

Barycentric subdivision for us will be a certain map

$$b_X: \|SX\| \longrightarrow \|SX\|$$

defined (for any space X) as follows: If $[\sigma,t]$ is the point of $\|SX\|$ represented by $(\sigma,t) \in S_nX \times \Delta^n$ then we choose an affine (injective) simplex $u: \Delta^q \longrightarrow \Delta^n$ belonging to the ordinary barycentric subdivision of Δ^n such that t is in the image of u. (More precisely: There is a strictly ascending sequence of q+1 faces of Δ^n such that u sends the vertices of Δ^q into the barycenters of these faces preserving the order.) Then we define

$$b_X[\sigma,t] = [\sigma u, u^{-1}t] .$$

It is easy to check that this neither depends on the choice of u nor on the choice of the representative (σ,t) of $[\sigma,t]$. (For the latter one uses the compatibility of barycentric subdivision with injective simplicial maps. Such a compatibility does not hold for "degeneracy maps" $\Delta^q \longrightarrow \Delta^p$, $p < q$, and this is the reason why we use the modified realization $\|\ \|$ at this point and not $|\ |$.)

We also need a homotopy

$$h_X\colon \|SX\| \times I \longrightarrow \|SX\|$$

from the identity of $\|SX\|$ to b_X. This is defined in a similar way as b_X : We consider the simplicial subdivision of $\Delta^n \times I$ whose vertices are the vertices of $\Delta^n \times 0$ and the barycenters of faces of $\Delta^n \times 1$. More precisely: Let $e_o,\dots,e_n$ be the vertices of Δ^n and let $b_{i_o\dots i_q}$ be the barycenter of the face of Δ^n spanned by $e_{i_o},\dots,e_{i_q}$. Then we take the set

$$V = \{(e_i,0)\,|\,i = 0,\dots,n\}$$
$$\cup\{(b_{i_o i_1\dots i_q},1)\,|\,0 \leq i_o < i_1 < \dots i_q \leq n\}$$

and give it the following partial order

$$(e_i,0) \leq (e_j,0) \quad \text{if} \quad i \leq j$$

$$(b_{i_o\dots i_q},1) \leq (b_{j_o\dots j_p},1) \quad \text{if} \quad \{i_o,\dots,i_q\} \subset \{j_o,\dots,j_p\}$$

$$(e_i,0) \leq (b_{i_o\dots i_q},1) \quad \text{if} \quad i \leq i_o .$$

The totally ordered subsets of V form our simplicial subdivision of $\Delta^n \times I$.

Now if $(\sigma,t) \in S_nX \times \Delta^n$ and $s \in I$ then we choose a simplex $u\colon \Delta^q \longrightarrow \Delta^n \times I$ belonging to the subdivision just defined (i.e. an affine map whose restriction to the set of vertices is a strictly increasing map into V) and such that (t,s) is in the image of u. We define

$$h_X([\sigma,t],s) = [\sigma\pi u, u^{-1}(t,s)] \quad ,$$

where $\pi\colon \Delta^n \times I \longrightarrow \Delta^n$ is the projection. Again it is easy to check that this neither depends on the choice of u nor on the choice of the representative (σ,t) of $[\sigma,t]$, and it is also easy to check that h_X is indeed a homotopy starting at the identity of $\|SX\|$ and ending at b_X .

It is clear that b_X and h_X are both natural with respect to continuous maps of X . In particular, returning to the hypotheses of the lemma, they are natural with respect to the inclusions $U \subset Y$, $U \in \mathcal{U}$. Hence b_Y is a map of the pair $(\|SY\|, \|S(Y;\mathcal{U})\|)$ into itself and h_Y is a homotopy of such maps.

If we fix for the moment one singular simplex $\sigma: \Delta^q \longrightarrow Y$, then by compactness and standard arguments about the size of simplices in the barycentric subdivision of Δ^q there is a natural number k such that each simplex of the k-th barycentric subdivision of Δ^q lies entirely in $\sigma^{-1}U$ for some $U \in \mathcal{U}$. This means that the k-th iterate of b_Y maps $[\sigma,t]$ into $\|S(Y;\mathcal{U})\|$ for each $t \in \Delta^q$.

Since $\|SY\|$ is a CW-complex (canonically) and $\|S(Y;\mathcal{U})\|$ is a subcomplex the lemma will be proved if we show that for each n any map of pairs

$$(D^n, S^{n-1}) \longrightarrow (\|SY\|, \|S(Y;\mathcal{U})\|)$$

(D^n the n-disc) is homotopic (as a map of pairs) to a map which sends all of D^n into the smaller space $\|S(Y;\mathcal{U})\|$ (cf. e.g. [W, II. 3.1 and 3.12]). By compactness $f(D^n)$ is contained in the union of finitely many closed cells of $\|SY\|$ which are nothing else but sets of the form

$$\{[\sigma,t] \mid t \in \Delta^q\} \quad , \quad \sigma: \Delta^q \longrightarrow X \text{ fixed.}$$

As we just remarked there is a natural number k such that $(b_Y)^k$ maps each of these cells and hence $f(D^n)$ into $\|S(Y;\mathcal{U})\|$.

Since b_Y is homotopic to the identity of $(\|SY\|, \|S(Y;\mathcal{U})\|)$ as a map of pairs so is the iterate $(b_Y)^k$. Hence f is homotopic to $(b_Y)^k f$ as a map of pairs, and the latter sends all of D^n into $\|S(Y;\mathcal{U})\|$. This proves the lemma.

3. Concluding remarks

From the results of Section 2 one obtains easily

3.1. COROLLARY. *The map* $\varepsilon_X\colon |SX| \longrightarrow X$ *is a weak homotopy equivalence for any space* X .

This follows from 2.2 and the following purely categorical observation.

3.2. PROPOSITION. *Let* (T,δ,ε) *be a comonad over some category* $\mathcal{D}$. *Then for all objects* A,X *of* $\mathcal{D}$ *the map*

$$\varepsilon_{X*}\colon \mathcal{D}(TA,TX) \longrightarrow \mathcal{D}(TA,X)$$

is surjective. It is injective if both ε_{TA} *and* ε_{TX} *are isomorphisms.*

Proof of 3.1. Let $\mathcal{D}$ be the homotopy category of topological spaces and (T,δ,ε) the comonad associated to the pair of adjoint functors

$$K \longmapsto |K|, \quad K \text{ a simplicial set}$$
$$X \longmapsto SX, \quad X \text{ a space}$$

[Mac, VI]. Then $TX = |SX|$. (Strictly speaking we get first a comonad on *Top* , but it induces a comonad on the homotopy category because T preserves homotopies, cf. 2.1 proof of condition (a).) The counit of the comonad is just (the homotopy class of) the map $\varepsilon_X\colon |SX| \longrightarrow X$ considered above (whereas we do not care what the diagonal is in our concrete situation). Thus we know from 2.2 that ε_{TA} is an isomorphism in $\mathcal{D}$ for all spaces A (because $TA \in \mathcal{W}$). Hence 3.2 implies that ε_X induces a bijection of homotopy sets

$$\varepsilon_{X*}\colon [TA,|SX|] \longrightarrow [TA,X]$$

for all spaces A,X . If $A \in \mathcal{W}$ then $\varepsilon_A\colon TA \cong A$ and we may replace TA by A .

Proof of 3.2. One of the axioms of comonads says that

$$\begin{array}{ccccc} & & TX & & \\ & \swarrow^{id} & \downarrow{\delta_X} & \searrow^{id} & \\ TX & \xleftarrow{\varepsilon_{TX}} & T^2X & \xrightarrow{T\varepsilon_X} & TX \end{array}$$

commutes for all objects X of $\mathcal{D}$. Using the left hand triangle (for A instead of X) and the fact that ε is a natural transformation one verifies that the composition

$$\mathcal{D}(TA,X) \xrightarrow{T} \mathcal{D}(T^2A,TX) \xrightarrow{\delta_A^*} \mathcal{D}(TA,TX) \xrightarrow{\varepsilon_{X*}} \mathcal{D}(TA,X)$$

is the identity which proves the surjectivity of ε_{X*} .

Assume now that ε_{TX} is an isomorphism. Then the above diagram shows that δ_X and $T\varepsilon_X$ are also isomorphisms and that δ_X is inverse to both ε_{TX} and $T\varepsilon_X$, hence $\varepsilon_{TX} = T\varepsilon_X$. We claim that the triangle

$$\begin{array}{ccc} \mathcal{D}(TA,TX) & \xrightarrow{\varepsilon_{X*}} & \mathcal{D}(TA,X) \\ & \searrow^{\varepsilon_{TA}^*} & \downarrow{T} \\ & & \mathcal{D}(T^2A,TX) \end{array}$$

commutes. To see this let $f \in \mathcal{D}(TA,TX)$. Then

$$\begin{array}{ccc} T^2A & \xrightarrow{Tf} & T^2X \\ \varepsilon_{TA}\downarrow & & \downarrow \varepsilon_{TX} = T\varepsilon_X \\ TA & \xrightarrow{f} & TX \end{array}$$

and we have indeed

$$\varepsilon_{TA}^* f = f\varepsilon_{TA} = (T\varepsilon_X)(Tf) = T(\varepsilon_X f) = T\varepsilon_{X*}f \ .$$

If also ε_{TA} is an isomorphism then ε_{TA}^* is bijective and hence ε_{X*} injective.

3.3. REMARK. Traditionally (except in [G]), before looking at the counit

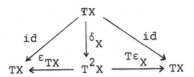

$$\varepsilon_X : |SX| \longrightarrow X$$

of the adjoint pair

$$K \longmapsto |K| \quad , \quad K \text{ a simplicial set}$$
$$X \longmapsto SX \quad , \quad X \text{ a space}$$

one looks at the unit

$$\eta_K : K \longrightarrow S|K|$$

and proves that it is a (weak) homotopy equivalence of simplicial sets. We get this result from 2.2 and the commutativity of

$$\begin{array}{ccc} |K| & \xrightarrow{|\eta_K|} & |S|K|| \\ & \searrow^{id_{|K|}} & \downarrow \varepsilon_{|K|} \\ & & |K| \end{array}$$

REFERENCES

[B] R.BROWN, Elements of modern topology, McGraw-Hill London 1968.

[D1] T.tom DIECK, Partitions of unity in homotopy theory, *Compos.Math.* 23 (1971), 159-167.

[D2] T.tom DIECK, On the homotopy type of classifying spaces, *manuscripta math.* 11 (1974), 41-49.

[G] J.B.GIEVER, On the equivalence of two singular homology theories, *Ann. of Math.* 51 (1950), 178-191.

[G-Z] P.GABRIEL and M.ZISMAN, *Calculus of fractions and homotopy theory*, Ergebnisse der Mathematik und ihrer Grenzgebiete 35, Springer 1967.

[L] K.LAMOTKE, *Semisimpliziale algebraische Topologie*, Die Grundlehren der mathematischen Wissenschaften in Einzeldarstellungen 147, Springer 1968.

[Mac] S.MAC LANE, *Categories for the working mathematician*, Graduate Texts in Mathematics 5, Springer 1971.

[May] J.P.MAY, *Simplicial objects in algebraic topology*, van Nostrand Mathematical Studies 11, van Nostrand 1967.

[Mi] J.MILNOR, The geometric realization of a semi-simplicial complex, *Ann. of Math.* 65 (1957), 357-362.

[P] D.PUPPE, Some well known weak homotopy equivalences are genuine homotopy equivalences, *Symposia Mathematica* 5 (1971), 363-374.

[V] R.M.VOGT, Homotopy limits and colimits, *Math.Z.* 134 (1973), 11-52.

[W] G.W.WHITEHEAD, *Elements of homotopy theory*, Graduate Texts in Mathematics 61, Springer 1978.

BETTI NUMBERS OF HILBERT MODULAR VARIETIES

E. Thomas
Department of Mathematics
University of California
Berkeley, CA 94720

A.T. Vasquez
Graduate School, CUNY
33 West 42nd Street
New York, NY 10036

1 INTRODUCTION

In this paper we give formulae for the Betti numbers of Hilbert modular varieties, as well as show that any such variety is simply-connected. These results complement the work of [19], where we calculate the Chern numbers of modular varieties of complex dimension three. The goal is the classification of modular varieties up to diffeomorphism, birational equivalence, or biholomorphic isomorphism; see section three for further discussion.

Throughout the paper K will denote a totally real algebraic number field of degree n (>1), $\mathcal{O}$ its ring of integers, and G $(= PSL_2(\mathcal{O}))$ its Hilbert modular group. We will say that a group Γ is of modular type (for K) if either $\Gamma = G$ or Γ is a torsion free subgroup of G of finite index. If Γ is a principal congruence subgroup (or $\Gamma = G$), we say that Γ is of principal type.

The group G acts on H (the complex upper half plane) by linear fractional transformations. Thus, by means of the n distinct embeddings of K in the real numbers, we obtain an action of G (and hence of Γ) on H^n. We define $Y_\Gamma = H^n/\Gamma$, the orbit space of this action. Y_Γ is a non-compact normal complex space with a finite number of isolated ("quotient") singularities, the images of the elliptic fixed points of the action of Γ on H^n.

Let h denote the number of parabolic orbits of Γ. (If $\Gamma = G$, h is simply the class number of K.) By adjoining h points (called "cusps") to Y_Γ one obtains a compact projective algebraic variety $\hat{Y}_\Gamma$ with isolated singularities: the cusps and the quotient singularities. See [11] for a detailed treatment of this material. We call any non-singular model Z_Γ of $\hat{Y}_\Gamma$ a Hilbert modular variety for Γ.

Our first result is:

THEOREM 1. *Let* K *be a totally real number field,* Γ *a group of principal type and* Z_Γ *a modular variety for* Γ. *Then,* Z_Γ *is simply-connected.*

This has been proved for quadratic number fields by Svarcman [16]. Moreover, he shows that for all n the singular varieties $\hat{Y}_\Gamma$ are simply-connected.

Note that by the theorem $b_1(Z_\Gamma) = 0$. (Here $b_i(X)$ denotes the ith Betti number of a space X). In the following section we give (computable) formulae for all Betti numbers of modular varieties, e.g., see Tables 3.2 and 3.4.

If $\deg K = 3$, we have an especially simple formula for b_3, at least for certain modular varieties. Recall that Z_Γ is obtained from $\hat{Y}_\Gamma$ by resolving the cusp and quotient singularities. Ehlers [3] gives an explicit method for resolving cusp singularities, while in [19] we give resolutions for the seven types of quotient singularities that arise for cubic number fields. If Γ is a group of modular type for a cubic number field K, we will say that a modular variety Z_Γ is special if:

(i) the cusp singularities in $\hat{Y}_\Gamma$ are resolved using Ehlers' construction;

(ii) the quotient singularities of $\hat{Y}_\Gamma$ are resolved using the models given in §2 of [19].

We then have:

THEOREM 2. *Let* K *be a totally real cubic number field,* Γ *a group of modular type for* K *and* Z_Γ *a special Hilbert modular variety for* Γ. *Then,*

$$b_3(Z_\Gamma) = 8(1 - \chi(\Gamma)) .$$

Here $\chi(\Gamma) = \chi(Z_\Gamma)$ = arithmetic genus of Z_Γ. (Since the arithmetic genus is a birational invariant (see §4), $\chi(\Gamma)$ depends only on Γ.)

Using a result of Freitag we may write b_3 in a different way. For each positive integer i set

$S_i(\Gamma)$ = dimension of the complex vector space of cusp forms (for Γ) of weight $2i$.

Freitag shows (7.2 in [4]): $S_1 = (-1)^n(\chi(\Gamma) - 1)$, where $\deg K = n$. Thus, by Theorem 2, we obtain:

COROLLARY 1. $b_3(Z_\Gamma) = 8S_1(\Gamma)$.

In the next section we give some evidence for the following conjecture (beyond the fact that $8 = 2^3$!).

CONJECTURE 1. *Let* K *be a totally real algebraic number field of odd degree* n , *with* Γ *a group of modular type for* K . *Then there is a Hilbert modular variety* Z_Γ *such that:*

$$b_n(Z_\Gamma) = 2^n S_1(\Gamma) .$$

2 BETTI NUMBERS

In this section we give formulae for the Betti numbers of modular varieties. Suppose that $\hat{Y}_\Gamma$ has h cusp singularities and s quotient singularities; let E_i, $1 \leqslant i \leqslant h$ and Q_j, $1 \leqslant j \leqslant s$, denote resolutions of these respective singularities. Topologically, we regard the E's and Q's as manifolds with boundary, and write:

$$Z_\Gamma = Y_\Gamma^0 \underset{\partial E_1}{\cup} E_1 \cup \cdots \underset{\partial E_h}{\cup} E_h \underset{\partial Q_1}{\cup} Q_1 \cup \cdots \underset{\partial Q_s}{\cup} Q_s .$$

Thus, Y_Γ^0 is a manifold with boundary, namely $\partial(\mathcal{Q}) \amalg \partial(\mathcal{E})$, where

$$\mathcal{E} = \coprod_i E_i , \qquad \mathcal{Q} = \coprod_j Q_j .$$

Since the E's and Q's are all disjoint we have for $k \geqslant 1$,

$$b_k(\mathcal{E}) = \sum_i b_k(E_i) , \qquad b_k(\mathcal{Q}) = \sum_j b_k(Q_j) .$$

We prove:

THEOREM 3. *Let* K *be a totally real algebraic number field of degree* n , Γ *a group of modular type for* K *and* Z_Γ *a Hilbert modular variety for* Γ . *Then, for* $1 \leqslant k \leqslant n - 1$,

$$b_k(Z_\Gamma) = b_k(\mathcal{Q}) + b_{2n-k}(\mathcal{E}) + \delta(n, k) ,$$

where

$$\delta(n, i) = \begin{cases} 0 , & \textit{if} \quad i \textit{ odd} ; \\ \binom{n}{i/2} , & \textit{if} \quad i \textit{ even} ; \end{cases} \qquad i \geq 0 .$$

The proof relies on work of Harder and is given at the end of the section.

Suppose now that x is a cusp singularity in $\hat{Y}_\Gamma$. As shown in [11], x is characterized by a pair (M, V), where M is a free Z-module of rank n in K and V is a group of totally positive units of rank $n - 1$. To construct a resolution E of x by Ehlers' method, one starts with a "complex of simplicies" Σ, defined using data from M and V. The group V then acts "simplicially" on Σ. (For details, see §1 in [3]; also, §1 in [18] for a brief resume of the method.) For $1 \leq i \leq n$, we set

$$N_i(E) = \mathrm{card}(\Sigma^{(i)}/V) ,$$

where $\Sigma^{(i)}$ denotes the set of i-simplices in Σ. It follows readily from [3] that $b_{2n-2}(E) = N_1(E)$, and so $b_{2n-2}(\mathcal{E}) = N_1(\mathcal{E}) = \sum N_1(E_i)$. Thus, as a corollary to Theorem 3 we obtain (if the E_i's are constructed by Ehlers' method):

$$b_2(Z_\Gamma) = b_2(\mathcal{Q}) + N_1(\mathcal{E}) + n . \tag{2.1}$$

In the following section we give some concrete examples.

Since Z_Γ has real dimension $2n$, by Theorem 3 and Poincaré duality we are left with only the calculation of $b_n(Z_\Gamma)$. Of course, by definition of the Euler characteristic, e, we have

$$b_n(Z_\Gamma) = (-1)^{n+1}[2 \sum_{i=1}^{n-1} (-1)^i b_i(Z_\Gamma) - e(Z_\Gamma) + 2] . \tag{2.2}$$

However, we wish to express b_n simply in terms of data from $\mathcal{E}$, $\mathcal{Q}$ and the field K, as in Theorem 3. For this we have:

THEOREM 4. *Let* K *be a totally algebraic number field of degree* n, *with* Z_Γ *a Hilbert modular variety. Then,*

$$b_n(Z_\Gamma) = (-1)^{n+1}(2\cdot\Delta(n) - e(Y_\Gamma)) + b_n(\mathcal{Q}) + b_n(\mathcal{E}) .$$

Here $\Delta(n) = \sum_{i=0}^{n-1} (-1)^i \delta(n, i)$. Moreover, by Hirzebruch [11],

$$e(Y_\Gamma) = 2[G : \Gamma]\zeta_K(-1) + \sum a_r(\Gamma)\frac{r-1}{r} .$$

Thus we have expressed b_n in the desired fashion.

Proof of Theorem 4. Since the E_i's and Q_j's have odd-dimensional boundaries, $e(Z_\Gamma) = e(\mathcal{Q}) + e(\mathcal{E}) + e(Y_\Gamma^0)$. But $e(Y_\Gamma) = e(Y_\Gamma^0) + s$, since $e(\text{disk}) = 1$. Using (2.2) and the fact that each ∂Q is a lens space, Theorem 4 now follows from Theorem 3 and the following fact.

FACT. *Let* $(E, \partial E)$ *denote any cusp resolution constructed by Ehlers' method. Then,*

$$e(E) = 2\sum_{j=0}^{n-1} (-1)^j b_{2n-j} + (-1)^n b_n ,$$

where $b_i = b_i(E)$.

We give the proof at the end of the section.

Suppose now that n is an odd integer (>1) , then $\Delta(n) = 2^{n-1}$; also, Vignéras [21] has shown that $e(Y_\Gamma) = 2^n\chi(\Gamma)$. Thus by Theorem 4 we have:

COROLLARY 2. *Suppose that* K *has odd degree* n . *Then,*

$$b_n(Z_\Gamma) = 2^n s_1(\Gamma) + b_n(\mathcal{Q}) + b_n(\mathcal{E}) .$$

Notice that by the corollary, Conjecture 1 follows from:

CONJECTURE 2. *Suppose that* K *has odd degree* n . *Then each quotient singularity resolution* Q *can be chosen so that*

(i) $b_n(Q) = 0$;

and each cusp resolution E *can be chosen so that*

(ii) $b_n(E) = 0$.

Proof of Theorem 2. Suppose that $\deg K = 3$ and that the resolution Q is chosen to be one of the seven resolutions constructed in §2 of [19]. Then, by Theorem 3.4 of [19], condition (i) given in Conjecture 2 is satisfied by Q. Thus, Theorem 2 follows at once from

LEMMA 1. *Let* K *be a totally real cubic number field and* E *a resolution of a cusp singularity for* K *constructed using an Ehlers' 3-complex. Then,* $b_3(E) = 0$.

We give the proof at the end of the section.

Proof of Theorem 3. We set

$$Z^0_\Gamma = Y^0_\Gamma \underset{\partial\mathcal{Q}}{\cup} \mathcal{Q} , \quad \text{so that} \quad \partial Z^0_\Gamma = \partial E .$$

Since each ∂Q_j is a lens space (and hence, rationally, a sphere) we see that

$$b_i(Z^0_\Gamma) = b_i(Y^0_\Gamma) + b_i(\mathcal{Q}) , \quad \text{for} \quad i > 0 . \tag{2.3}$$

Suppose now that E is a resolution of a cusp singularity, with ∂E as a boundary. Taking homology with complex coefficients we have:

CLAIM. *For* $1 \leqslant i \leqslant n - 1$, $H_i(\partial E) \longrightarrow H_i(E)$ *is injective.*

Assuming this for the moment we have:

Proof of Theorem 3. By the Claim, the following two sequences are exact, for $1 \leqslant i \leqslant n - 1$:

$$\text{(a)} \quad 0 \longrightarrow H_i(\partial E) \longrightarrow H_i(E) \longrightarrow H_i(E, \partial E) \longrightarrow 0 ,$$

$$\text{(b)} \quad 0 \longrightarrow H_i(\partial E) \longrightarrow H_i(Z^0_\Gamma) \oplus H_i(E) \longrightarrow H_i(Z_\Gamma) \longrightarrow 0 .$$

Therefore, by (a) and (b),

$$\begin{aligned} b_i(Z_\Gamma) &= b_i(Z^0_\Gamma) + b_i(E) - b_i(\partial E) \\ &= b_i(Z^0_\Gamma) + b_i(E, \partial E) . \end{aligned}$$

Consequently, by Lefschetz duality and (2.3) we have:

(c) $b_i(Z_\Gamma) = b_i(Y_\Gamma^0) + b_i(Q) + b_{2n-i}(E)$.

Denote by S the set of quotient singularities in Y_Γ . Thus, Y_Γ^0 has the homotopy type of $Y_\Gamma - S$; moreover,

$$b_i(Y_\Gamma^0) = b_i(Y_\Gamma - S) = b_i(Y_\Gamma) , \quad 0 \leq i \leq 2n - 2 .$$

Thus the proof of Theorem 3 follows at once from (c), when we show:

(d) For $1 \leq i \leq n - 1$, $b_i(Y_\Gamma) = \delta(n, i)$.

Suppose first that Γ is torsion free. Then (d) follows from the work of Harder [8] (see also [5]). By Theorem 2.1(i), Remark (ii) on page 145, and the reference to the work of Matsushima and Shimura given on page 147 (all references are to [8]), we see that

$$H^i(Y_\Gamma) = H_A^i(Y_\Gamma) , \quad 1 \leq i \leq n - 1 \quad \text{complex coefficients}$$

where the right hand group is isomorphic to the elements of degree i in the exterior algebra generated by certain 2-forms on Y_Γ , $\eta_1, \ldots, \eta_n$ (op. cit., page 146). Thus, $\dim H_A^i(Y_\Gamma) = \delta(n, i)$, which proves (d) and hence Theorem 3, in this case.

On the other hand, suppose that Γ has torsion; i.e., $\Gamma = G$. Let Γ' be a torsion free subgroup of finite index -- e.g., take Γ' to be an appropriate principal congruence subgroup. Set $F = \Gamma/\Gamma'$, so that F acts on $Y_{\Gamma'}$, with $Y = Y_{\Gamma'}/F$. Thus, by Theorem III.2.4 of [8],

$$H^*(Y_\Gamma) \approx H^*(Y_{\Gamma'})^F = \text{submodule of } H^*(Y_{\Gamma'}) \text{ left invariant by } F .$$

As observed above, $H^i(Y_{\Gamma'}) \approx H_A^i(Y_{\Gamma'})$, for $1 \leq i \leq n - 1$, and so $H^i(Y_\Gamma) \approx H_A^i(Y_{\Gamma'})^F$. But Harder [8] shows that the 2-forms $\eta_1, \ldots, \eta_n$ (see above) are all defined back on H^n , where they are in fact $PSL_2(R)^n$-invariant. Hence, each η_i is G-invariant and so (on $Y_{\Gamma'}$), F-invariant. Thus, $H_A^i(Y_{\Gamma'})^F = H_A^i(Y_{\Gamma'})$, which implies $\dim H^i(Y_\Gamma) = \delta(n, i)$, as before. This completes the proof of the theorem.

Proof of Claim. It clearly suffices to prove the analogous result:

(e) For $1 \leqslant i \leqslant n-1$, $H^i(E) \longrightarrow H^i(\partial E)$ is onto.

As shown in [11], ∂E fibers over the $n-1$ torus, say $p : \partial E \longrightarrow T^{n-1}$. By Proposition 1.1 in [8], $p^* : H^i(T^{n-1}) \approx H^i(\partial E)$, for $1 \leqslant i \leqslant n-1$. Since $H^*(T^{n-1})$ is generated by 1-dimensional classes, it suffices to prove (e) simply in the case $i = 1$. For this we need the following result, which will be proved in section 5.

LEMMA 2. $b_1(E) = n - 1$.

Thus $H^1(E)$ and $H^1(\partial E)$ are both complex vector spaces of the same dimension, and so to prove (e) (with $i = 1$), it suffices to show that $H^1(E) \longrightarrow H^1(\partial E)$ is injective. This follows by exactness if we know that $H^1(E, \partial E) = 0$. But by Lefschetz duality, $H^1(E, \partial E) \approx H_{2n-1}(E)$, which is zero since E has the homotopy type of the singular set (of dimension $2n - 2$) . This completes the proof of the Claim.

Proof of Fact. By the exactness of sequence (a) above, and by Lefschetz duality,

$$b_j = b_{2n-j} + c_j \quad , \qquad 1 \leqslant j \leqslant n-1 \ ,$$

where $c_j = b_j(\partial E)$. Thus,

$$e(E) = \sum_{j=0}^{n-1} (-1)^j b_j + (-1)^n b_n + \sum_{j=0}^{n-1} (-1)^j b_{2n-j}$$

$$= 2 \cdot \sum_{j=0}^{n-1} (-1)^j b_{2n-j} + (-1)^n b_n + \sum_{j=0}^{n-1} (-1)^j c_j \ .$$

But as shown above, in the proof of the Claim,

$$c_1 = b_i(T^{n-1}) \quad , \qquad 1 \leqslant i \leqslant n-1 \ ,$$

where T^{n-1} is an $(n - 1)$-torus. Thus,

$$\sum_{j=0}^{n-1} (-1)^j c_j = e(T^{n-1}) = 0 \ .$$

This completes the proof.

Proof of Lemma 1. We use the notation from [3] and [17]. Thus, we assume that the resolution E is given by a pair (M, V) , with $E = X/V$ for a certain (open) complex n-manifold on which the group

V acts freely and properly discontinuously. To prove the lemma we will find an appropriate subgroup V' of finite index k in V such that:

(i) $N_j(E') = kN_j(E)$, where $E' = X/V'$, $j > 0$;
(ii) $2N_1(E') = N_3(E')$.

Let C denote the set of points (x_1, x_2, x_3) in R^3 such that each $x_i > 0$ and $x_1x_2x_3 = 1$. We regard K (and hence V) as embedded in R^3 by the three real embeddings of K; also, V acts on R^3 by coordinate-wise multiplication. With this action, C is then V-invariant. Let $\ell : C \longrightarrow R^3$ by

$$(x_1, x_2, x_3) \longrightarrow (\log x_1, \log x_2, \log x_3) .$$

By ℓ, C is homeomorphic to the plane $\Pi : x_1 + x_2 + x_3 = 0$ in R^3, while V is isomorphic to a lattice of rank two in Π. Thus,

$$C/V \approx \Pi/Z^2 = T^2 , \quad \text{a two-dimensional torus.}$$

Note that the Ehlers' complex Σ, which gives rise to the manifold X, induces a triangulation on the space C (see pp. 7-8 in [17]); in particular, each $\sigma \in \Sigma^{(j)}$ (= set of j-simplices in Σ) gives a $(j - 1)$-simplex on C. The action of V is then simplicial with respect to this triangulation. Thus C/V inherits a cell structure, with $N_j(E)$ = number of $(j - 1)$-cells on C/V. Similar remarks obtain for C/V', where V' is any group of finite index in V. Since $C/V = (C/V')/F$, where F $(= V/V')$ acts freely, it follows that

$$k = |F| = \text{number of cells of } C/V' \text{ lying over each cell of } C/V .$$

Thus, $N_j(E') = kN_j(E)$, as asserted above.

On the other hand, since $C/V' = T^2$,

$$e(C/V') = 0 = N_1(E') - N_2(E') + N_3(E') .$$

It follows readily from Lemma 3.4 in [18] that there is a V' as above such that the cell structure on C/V' is actually a simplicial triangulation. Since each triangle has 3 edges and each edge lies on precisely 2 triangles, $3N_3(E') = 2N_2(E')$. Consequently,

$$2N_1(E') = 2N_2(E') - 2N_3(E') = N_3(E') .$$

This proves (ii) above, and hence shows that $2N_1(E) = N_3(E)$. On the other hand, by the above Fact,

$$b_3 = 2b_4 - e(E) = 2N_1 - N_3 = 0 ,$$

since $e(E) = N_3$ by Theorem 12 of [3]. This completes the proof of Lemma 1.

REMARK. Theorems 3 and 4 hold even if Γ has torsion.

3 BETTI NUMBERS AND HODGE NUMBERS

We illustrate the preceding material with two sets of examples, both coming from cubic number fields. Let K be such a (totally real) field with Γ a group of modular type. In [19], for any special modular variety Z_Γ for Γ, we have computed the Chern numbers χ, e, and c_1^3. Since

$$b_2 = \tfrac{1}{2}(e + b_3) - 1 , \qquad (3.1)$$

$b_2(Z_\Gamma)$ can be readily calculated from e and χ, by Theorem 2. Also, by Theorem 3, $b_1(Z_\Gamma) = 0$, since E has the homotopy type of a 4-complex and $b_1(Q) = 0$ by (3.4) of [19].

Consider the family of Galois cubic number fields $K = \mathbb{Q}(\lambda)$, where

$$\lambda^3 + (n + 1)\lambda^2 + (n - 2)\lambda - 1 = 0 , \qquad n > 0 .$$

For $n = 1, 2, 3, 4, 6, 9, 10$ or 12 these fields have class number 1 [6] and hence Y_G has but a single cusp (G denotes the Hilbert modular group for K) . In [17] we give an explicit resolution for this cusp, while in [19] we give explicit resolutions for all quotient singularities that arise for cubic number fields. Let $Z(f)$ denote the resulting modular variety for G_K, where

$$f = \text{conductor } K = n(n - 1) + 7 .$$

In the following table we give the Chern numbers for these eight varieties, and then calculate the Betti numbers b_2 and b_3 as indicated above. We also calculate the Hodge numbers $h^{3,0}$ and $h^{2,1}$.

The Hodge numbers $h^{p,q}$ are calculated as follows. Since Z_Γ is Moishezon (it dominates $\hat{Y}_\Gamma$, a compact projective variety), $h^{p,q} = h^{q,p}$ and $b_i = \sum_{p+q=i} h^{p,q}$ [2]. Also, Freitag [5] shows that $h^{2,0}(Z_\Gamma) = 0$; and $h^{1,0} = 0$ since $b_1 = 0$. Thus,

$$h^{1,1} = b_2 , \quad h^{3,0} = h^{0,3} = 1 - \chi = b_3/8 , \quad h^{2,1} = 3b_3/8 \qquad (3.2)$$

TABLE: Betti Numbers and Hodge Numbers of $Z(f)$ (3.3)

f	χ	c_1^3	e	b_2	b_3	$h^{3,0}$	$h^{2,1}$
7	1	-6	64	31	0	0	0
9	1	-6	74	36	0	0	0
13	1	-4	56	27	0	0	0
19	1	-6	74	36	0	0	0
37	0	-64	100	53	8	1	3
79	-11	-762	190	142	96	12	36
97	-24	-1420	156	177	200	25	75
139	-82	-4218	-234	214	664	83	249

If X is a rational projective variety, then $\chi(X) = 1$. Thus, for $f = 7, 9, 13$, or 19 the varieties $Z(f)$ are possibly rational. (It is shown in [22] that if conductor $K > 19$, then $\chi(G_K) \neq 1$.) Thus (as noted in [22]) we have:

Problem 1. For $f = 7, 9, 13$, or 19 is the Hilbert modular variety $Z(f)$ rational? If not, is it unirational?

By [12], for $f = 37, 79, 97$ and 139 the variety $Z(f)$ is of general type.

Note also from the Table that $Z(9)$ and $Z(19)$ have precisely the same invariants. Thus, we have:

Problem 2. Are $Z(9)$ and $Z(19)$ diffeomorphic? If so, are they biholomorphically isomorphic?

In the study of Hilbert modular surfaces it has proved useful to study surfaces Z_Γ where Γ is a torsion free principal congruence subgroup of G. We now calculate the invariants given in (3.3) for three such examples, K cubic. Here $\Gamma(P)$ denotes the (torsion free) principal congruence subgroup associated to a prime ideal P ($= (\lambda - r)$), see (1.7) in [19].

TABLE: Betti Numbers and Hodge Numbers for $Z_{\Gamma(P)}$ (3.4)

f	P	χ	c_1^3	e	b_2	b_3	$h^{3,0}$	$h^{2,1}$
7	$(\lambda + 3)$	-2	-48	80	51	24	3	9
13	$(\lambda + 2)$	-5	-144	80	63	48	6	18
19	$(\lambda + 3)$	-165	-7080	-360	483	1328	166	498

4 BIRATIONAL ISOMORPHISM

If V and V' are complex projective varieties, one has the classical notion of V and V' being birationally isomorphic (or just "birational") -- e.g., see [9]. Since we wish to consider modular varieties that are not necessarily projective it is convenient to extend slightly this notion. If V and V' are just complex spaces, Remmert has introduced the important idea of V and V' being bimeromorphically equivalent [14]. Suppose that V and V' are compact non-singular complex manifolds. We will then call V and V' birationally isomorphic if they are bimeromorphically equivalent. considered simply as complex spaces (compare [7], page 493).

Let Z_Γ be a Hilbert modular variety for Γ, a group of modular type. Thus there is a map p from Z_Γ to $\hat{Y}_\Gamma$ such that p is a biholomorphic isomorphism from $Z_\Gamma - p^{-1}(S)$ to $\hat{Y}_\Gamma - S$, where S is the set of singular points in $\hat{Y}_\Gamma$. Moreover Z_Γ is compact and non-singular. Thus, if Z'_Γ is a second model for $\hat{Y}_\Gamma$, Z_Γ and Z'_Γ are birationally isomorphic. (See pages 13-17 of [20]). In particular,

$$\chi(Z_\Gamma) = \chi(Z'_\Gamma) \tag{4.1}$$

by Corollary 2.15 in [20]. (Following Hirzebruch [10], for any complex n-manifold X we set

$$\text{arithmetic genus } X = \chi(X) = \sum_{i=0}^{n} (-1)^i \dim H^i(X; \Omega_X) ,$$

where Ω_X is the structure sheaf for X).

For use in §5 we show (compare, [7] page 494).

PROPOSITION 4.2. *Let* Z *and* Z' *be two non-singular models for* $\hat{Y}_\Gamma$. *Then,*

$$\pi_1(Z, *) \approx \pi_1(Z', *) .$$

Proof. As noted above, Z and Z' are birationally isomorphic. Thus (see [20]) there are open sets

$$U \subset V \overset{i}{\subset} Z \ , \qquad U' \subset V' \overset{i'}{\subset} Z'$$

and (holomorphic) maps

$$f : V \longrightarrow Z' \ , \qquad f' : V' \longrightarrow Z$$

such that

$$\operatorname{codim}(Z - U) \geqslant 1 \ , \qquad \operatorname{codim}(Z - V) \geqslant 2 \ ;$$

$$\operatorname{codim}(Z' - U') \geqslant 1 \ , \qquad \operatorname{codim}(Z' - V') \geqslant 2 \ ;$$

$$f(U) = U' \ , \quad f'(U') = U \ , \quad f|U = (f'|U')^{-1} : U \approx U' \ .$$

(In fact, $U = Z - p^{-1}(S)$, $U' = Z' - p'^{-1}(S)$) .

Therefore,

$$\pi_1(V, *) \xrightarrow{i_*} \pi_1(Z, *) \ , \qquad \pi_1(V', *) \xrightarrow{i'_*} \pi_1(Z', *) \ .$$

Set $\phi = f_* \circ i_*^{-1}$, $\phi' = f'_* \circ i'^{-1}_*$. Clearly, $\phi' \circ \phi = \mathrm{id}$ on $\pi_1(U, *)$ and $\phi \circ \phi' = \mathrm{id}$ on $\pi_1(U', *)$. Since

$$\pi_1(U, *) \longrightarrow \pi_1(Z, *) \quad \text{and} \quad \pi_1(U', *) \longrightarrow \pi_1(Z', *)$$

are onto it follows that $\phi' \circ \phi = \mathrm{id}$ on $\pi_1(Z, *)$ and $\phi \circ \phi' = \mathrm{id}$ on $\pi_1(Z', *)$. This completes the proof.

5 PROOF OF THEOREM 1

As noted in §4, the fundamental group is a birational invariant, and so we may assume that Z_Γ is constructed by Ehlers' method. Write Z_Γ as in (2.1) and set $Z_0 = Y_\Gamma^0$ so that Z_Γ is constructed from Z_0 in $h + s$ stages, adding first the successive E_i's and then the Q_j's . We take the first cusp to be ∞ ; thus Γ_∞ (the isotropy subgroup of ∞) $\subset G_\infty$, and so by page 235 of [11], Γ_∞ is (isomorphic to) the group of matrices

$$\begin{pmatrix} \varepsilon^2 & \mu \\ 0 & 1 \end{pmatrix} \qquad \varepsilon \equiv 1 \bmod A \ , \qquad \mu \in A \ ,$$

where ε is a unit in $\mathcal{O}$ and A is the integral ideal that defines the principal congruence subgroup Γ .

Set $Z_k = Z_{k-1} \cup P_k$, $1 \leqslant k \leqslant h + s$, where $\{P_k\} = \{E_1, \ldots, E_h, Q_1, \ldots, Q_s\}$. To prove the theorem it suffices to show:

(a) $\pi_1(Z_1, *) = 1$

(b) $\pi_1(Z_{k-1}, *) = 1 \Longrightarrow \pi_1(Z_k, *) = 1$, $k > 1$.

We begin with assertion (b). (For simplicity we now write simply $\pi(X)$ for $\pi_1(X, *)$) . Note that $P_k \cap Z_{k-1} = \partial P_k$; also, P_k contains in its interior the singular set (= exceptional divisor) S_k . Since $P_k - S_k$ has ∂P_k as a strong deformation retract and since S_k has real codimension 2 , $\pi(\partial P_k) \longrightarrow \pi(P_k)$ is onto. Thus, (b) follows at once from the Seifert-Van Kampen Theorem (e.g., [13]).

To prove (a), set $E = E_1$ and consider the diagram:

$$\text{(c)} \qquad \begin{array}{ccc} \pi(\partial E) & \xrightarrow{j_*} & \pi(E) \\ \downarrow{\scriptstyle i_*} & & \downarrow \\ \pi(Z_0) & \longrightarrow & \pi(Z_1) \end{array}$$

As above j_* is onto, and so by Seifert-Van Kampen,

(d) $\pi(Z_1) \approx \pi(Z_0)/\text{normal closure of } i_*(\text{Ker } j_*)$.

Note that $\partial E = R^{2n-1}/\Gamma_\infty$ (see page 194 [11]), where Γ_∞ acts freely, so $\pi(\partial E) = \Gamma_\infty$. Similarly, $\pi(Z_0) = \pi(Y^0_\Gamma) \approx \Gamma$. Moreover, we may identify $i_* : \pi(\partial E) \longrightarrow \pi(E)$ with the natural inclusion $\Gamma_\infty \subset \Gamma$; in particular, i_* is a monomorphism. Thus, by (d) we will prove Theorem 1 when we show:

$$\Gamma = \text{normal closure of } i_*(\text{Ker } j_*) \ .$$

Consequently, by a result of Serre (Cor. 3, page 499, in [15]), Theorem 1 is proved when we prove the following result.

Let E be a resolution of any cusp (M, V) , as above, so that $\partial E = R^{2n-1}/\Gamma(M, V)$, where $\Gamma(M, V)$ is the group of matrices $\begin{pmatrix} v & m \\ 0 & 1 \end{pmatrix}$, with v in V and m in M . Thus, $\pi(\partial E) \approx \Gamma(M, V)$.

Regard M as a subgroup of $\Gamma(M, V)$ by the map

$$m \longmapsto \begin{pmatrix} 1 & m \\ 0 & 1 \end{pmatrix} \ .$$

LEMMA 3. *Let* j *denote the inclusion* $\partial E \hookrightarrow E$. *Then, with the above identifications,* $M \subset \operatorname{Ker} j_*$.

Proof. Recall that $\Gamma(M, V)$ acts on H^n by the formula:

$$\begin{pmatrix} v & m \\ 0 & 1 \end{pmatrix}(z_1, \ldots, z_n) = (\tilde{z}_1, \ldots, \tilde{z}_n) ,$$

$$\tilde{z}_j = v^{(j)} z_j + m^{(j)} .$$

Here $x^{(j)} \in R$ denotes the image of x in K by the jth imbedding $K \subset R$. We write $\Gamma = \Gamma(M, V)$, and denote by p', p'' and p the canonical projection maps

$$H^n \xrightarrow{p'} H^n/M \xrightarrow{p''} H^n/\Gamma , \qquad p = p'' \circ p' .$$

Define $\nu : H^n \longrightarrow R_+$ by

$$(z_1, \ldots, z_n) \longmapsto \operatorname{Im}(z_1) \cdots \operatorname{Im}(z_n) .$$

For any c in R_+ , $\partial B_c = \nu^{-1}(c)$ is Γ-stable, and for a fixed c we identify ∂E with $\partial B_c/\Gamma$, as above.

Let $b \in \partial B_c$; covering space theory gives an isomorphism (which depends on b)

$$\theta_b : \Gamma \approx \pi_1(\partial E, p(b)) .$$

We note for later use the special case, $\theta_b(m) = [p \circ \gamma_m]$ for $m \in M$, where $\gamma_m : [0, 1] \longrightarrow \partial B_c$ by

$$t \longmapsto b + tm \quad (= b^{(j)} + tm^{(j)} , \text{ in the } j^{th} \text{ coordinate}, \; 1 \leq j \leq n) .$$

Now let Σ be an Ehlers' n-complex constructed using M and V , X_Σ the associated complex n-manifold, and F_Σ the "exceptional divisor" in X_Σ (see §1 in [3]). There is a biholomorphic map $\pi : X_\Sigma - F_\Sigma \approx C^n/M$; and if we set $X = \pi^{-1}(H^n/M) \cup F_\Sigma$, $E = X/V$ is then the desired resolution for the cusp (M, V) . By an abuse of notation we also write E for the manifold with boundary, $\pi^{-1}(p'(B_c)) \cup F_\Sigma$, where $B_c = \{\nu^{-1}(d) \in H^n | c \leq d \in R_+\}$.

Let $\hat{b} = \pi^{-1}(p'(b))$. There is then an n-simplex $\sigma = \langle v_1, \ldots, v_n \rangle$ in Σ such that $\hat{b} \in (C^n)_\sigma \cap X$. Suppose that $\hat{b}$ has coordinates $(z_1, \ldots, z_n)$ in C^n_σ ; recall the formula for π on C^n_σ :

$$(z_1, \ldots, z_n) \longrightarrow p'(w_1, \ldots, w_n) \in C^n/M ,$$

where $w_j = \frac{1}{2\pi i} \sum_{s=1}^{n} \log(z_s) v_s^{(j)}$. Writing $z_s = r_s e^{i\phi_s}$, this becomes:

$$w_j = \sum_{s=1}^{n} \frac{\phi_s}{2\pi} v_s(j) + i \cdot \left(\sum_{s=1}^{n} \frac{-\log(r_s)}{2\pi} v_s^{(j)} \right) .$$

Define $\gamma_k : [0, 1] \longrightarrow C^n_\sigma$ by

$$t \longmapsto (z_1, \ldots, z_{k-1}, e^{2\pi i t} z_k, \ldots, z_n) , \qquad 1 \leqslant k \leqslant n .$$

The above formulae for π and ν show that $p'' \circ \pi \circ \gamma_k$ is a loop in $\partial B_c/\Gamma$ and the formula for the isomorphism θ_b shows that this loop represents $\theta_b(v_k)$ in $\pi_1(\partial E, p(b))$. Since $\{v_1, \ldots, v_n\}$ is a basis for M , to prove Lemma 3 it suffices to show that each loop γ_k can be deformed to a constant loop in $C^n_\sigma \cap X$. This deformation takes place in $(z_1, \ldots, z_{k-1}) \times C \times (z_{k+1}, \ldots, z_n)$, as suggested by the picture:

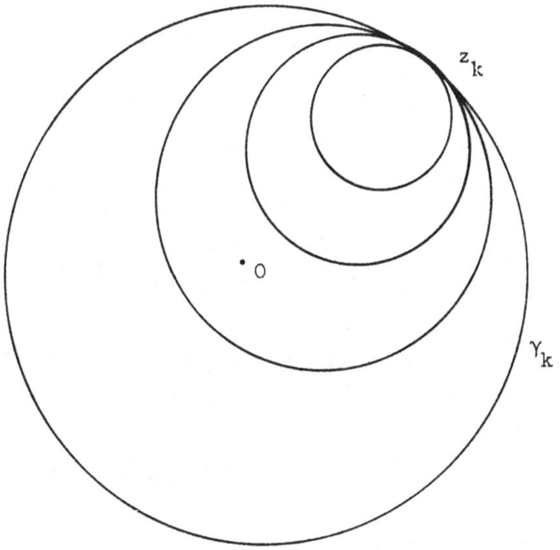

Figure 1

The formulae for π and ν show that this contraction in fact takes place in $C^n_\sigma \cap X$. (Recall that the $v_s^{(j)}$'s are all positive numbers and that if $(z_1, \ldots, z_n) \in C^n_\sigma \cap X$ then so does $(z_1, \ldots, z_{k-1}, w, z_{k+1}, \ldots, z_n)$ for all $w \in C$ with $0 \leqslant |w| \leqslant |z_k|$).

This completes the proof of Lemma 3 and hence of Theorem 1.

Proof of Lemma 2. As noted above $E = X/V$. Since V is a free Z-module of rank $n - 1$ which acts freely and properly discontinuously on X, Lemma 2 follows at once from:

CLAIM. X *is simply-connected.*

Proof. Let $V(\sigma) = X \cap (C^n)_\sigma$ for $\sigma \in \Sigma^{(n)}$. (Notation is that of Ehlers, [3], page 134). Order the σ's in $\Sigma^{(n)}$ so that if $X_j = \bigcup_{1 \leq i \leq j} V(\sigma_i)$, then

$$X = \bigcup_j X_j \quad \text{and} \quad X_j \cap V(\sigma_{j+1}) \neq \emptyset .$$

Since any compact subset of X is in X_j for j large enough, it suffices to show that each X_j is 1-connected. For this the Seifert-Van Kampen Theorem and the following Lemma suffice.

LEMMA 4. (i) $V(\sigma_k)$ *is contractible, all* k.
(ii) $V(\sigma_k) \cap X_{k-1}$ *is arcwise-connected, all* k.

The proof follows from the definition of the manifold X given in [3], page 134. We omit the details.

REMARK. We take this opportunity to correct several misprints in [19]. On page 194, line 15 should begin: $\ldots + (n + 1)\lambda^2 + (n - 2)\lambda \ldots$, while in line 21, [20] should be [21]. Also, the parenthetical sentence in line 19 should read: (...for all other n (with $F(n)$ not one of the above 8 fields), ...). Finally, on page 205, line 11, Figure 3 should be Figure 2

Research supported by grants from the National Science Foundation.

REFERENCES

[1] Bredon, G. (1972). Introduction to Compact Transformation Groups. New York: Academic Press.

[2] Deligne, P. (1971). Theorie de Hodge, II. Publ. Math. I.H.E.S., 40, 5-58.

[3] Ehlers, F. (1975). Eine Klasse komplexer Mannigfaltigeiten und die Auflösung einiger isolierter Singularitäten. Math. Ann., 218, 127-156.

[4] Freitag, E. (1972). Lokale und globale invarianten der Hilbertschen modulgruppe. Invent. Math., 17, 106-134.
[5] Freitag, E. (1975). Singularitäten von Modulmannigfaltigkeiten und Körper Automorpher Funktionen. Proc. Int. Congress Math., 443-448.
[6] Gras, M. (1975). Méthodes et algorithmes pour le calcul numerique du nombre de classes et des unités des extensions cubique cyclicques de Q. Journal f. d. r. u. a. Math. (Crelle), 227, 89-116.
[7] Griffiths, P. & Harris, J. (1978). Principles of Algebraic Geometry. New York: John Wiley & Sons.
[8] Harder, G. On the cohomology of SL(2, $\mathcal{O}$). In Lie Groups and Their Representations, 139-150.
[9] Hartshorne, R. (1977). Algebraic Geometry. In Graduate Texts in Mathematics, no. 52. New York: Springer-Verlag.
[10] Hirzebruch, F. (1966). Topological Methods in Algebraic Geometry, 3rd ed. Berlin: Springer-Verlag.
[11] Hirzebruch, F. (1974). Hilbert modular surfaces. L'Enseignement Math., 19, 183-281.
[12] Knöller, F.W. Beispiele dreidimensionaler Hilbertscher Modulmannigfaltigkeiten von allgemeinem Typ. To appear.
[13] Massey, W. (1977). Algebraic topology: an introduction. In Graduate Texts in Math., no. 56. Berlin: Springer-Verlag.
[14] Remmert, R. (1957). Holomorphe und meromorphe Abbildungen komplexer Raüme. Math. Annalen, 133, 328-330.
[15] Serre, J.-P. (1970). Le problème des groupes de congruence pour SL_2. Ann. Math., 92, no. 2, 489-527.
[16] Svarcman, O.V. (1974). Simple-connectivity of a factor-space of the Hilbert modular group. Functional Analysis and its Applications, 8, 99-100.
[17] Thomas, E. & Vasquez, A. (1980). On the resolution of cusp singularities and the Shintani decomposition in totally real cubic number fields. Math. Ann., 247, 1-20.
[18] Thomas, E. & Vasquez, A. (1981). Chern numbers of cusp resolutions in totally real cubic number fields. J. Reine Ang. Math., 324, 175-191.
[19] Thomas, E. & Vasquez, A. (1981). Chern numbers of Hilbert modular varieties. J. Reine Ang. Math., 324, 192-210.
[20] Ueono, K. (1975). Classification theory of algebraic varieties. In Lecture Notes in Math., no. 439. New York: Springer-Verlag.
[21] Vignéras, M.-F. (1976). Invariants numériques des groupes de Hilbert. Math. Ann., 224, 189-215.
[22] Weisser, D. The arithmetic genus of the Hilbert modular variety and the elliptic fixed points of the Hilbert modular group. Math. Ann. To appear.

HOMOTOPY PAIRS IN ECKMANN-HILTON DUALITY

K.A. Hardie

0. Introduction

In a long series of papers appearing over the seven years 1958-64 Eckmann and Hilton studied in depth a partly heuristic duality in the homotopy theory of pointed topological spaces. Hilton has recently written a retrospective essay [13] discussing their motivation and philosophy including, together with a survey of more recent developments, a rather complete list of papers having a bearing on the subject.

Basic to the duality was the concept of *pair*. Classically this derived from relative topology whose objects were indeed pairs of spaces (X, A) with $A \subset X$. However with every such pair is associated an inclusion *map* $A \to X$. On translation into categorical language the arrow is more important than its domain and codomain. Thus (and for other good reasons) a pair came to mean a (pointed) continuous map. In formulating the naturality properties of the two exact sequences in homotopy associated with a pair it was necessary to consider maps of pairs. From relative topology the obvious concept to take for a map $f \to g$ was a (strictly) commutative diagram

(0.1)
$$\begin{array}{ccc} A & \xrightarrow{\psi} & E \\ {\scriptstyle f}\downarrow & & \downarrow{\scriptstyle g} \\ X & \xrightarrow{\phi} & B \end{array}$$

and later when contemplating homotopy of maps of pairs it was natural to consider homotopies $\phi_t : X \to B$ and $\psi_t : A \to E$ satisfying $\phi_t f = g\psi_t$ for each $t \in I$. Yet, as has been discussed in [8], [9] and [10] the associated pair-homotopy category PC is not from all points of view the most convenient one to study. To mention two considerations only : the morphism sets $[f, g]$ in the pair-homotopy category are not always invariants of the homotopy classes of f and of g; the notion of homotopy equivalence of pairs that features heavily in the work on the Puppe sequence [18] is weaker than that of pair-homotopy equivalence. For these and other reasons the authors of the papers cited have considered homotopy commutative diagrams.

A *homotopy pair map* from f to g is a triple $(\phi, \psi, \{h_t\})$ where h_t is a homotopy from ϕf to $g\psi$ and $\{h_t\}$ is the associated *track* or homotopy class of homotopies from ϕf to $g\psi$. That this definition gives rise to a category HPM (in fact to a 2-category) was already observed by Gabriel-Zisman [3]. The set $\pi(f, g)$ of *homotopy pair classes* from f to g is obtained by factoring out by the relation

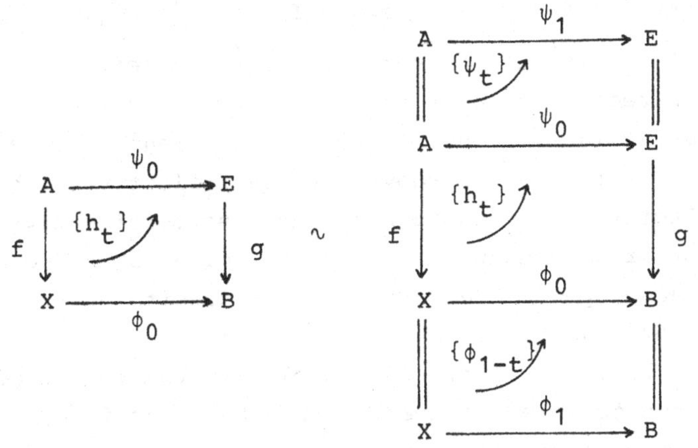

where the square referred to on the right is the composite in the obvious sense of the three internal squares. In the resulting category HPC homotopy equivalent pairs are isomorphic, the Puppe construction $f \rightsquigarrow Pf$ is functorial (Pf is the pair that includes the codomain of f into the cofibre of f) and P is left adjoint to N (Nf is the pair that maps the fibre of f into the domain of f) : for details see [9]. In similar vein every pair f is isomorphic in HPC to Jf (the inclusion of the domain of f into its mapping cylinder), thus every pair is isomorphic with a pointed cofibration.

Alternatively we may regard J as a functor from HPC to PC which then is a full and faithful left adjoint to the obvious functor $I : PC \to HPC$. For details see [8]. Dually every pair f is isomorphic in HPC to Lf (the projection of the mapping-track of f onto the codomain of f), thus every pair is isomorphic with a Hurewicz fibration. Moreover $L : HPC \to PC$ is a full and faithful right adjoint of I. Since homotopic maps are evidently isomorphic as pairs the homotopy invariance of $\pi(f, g)$ is assured. On the other hand the relationship between $\pi(f, g)$ and $[f, g]$ is clarified by observing that if f is an h-cofibration then f and Jf have the same pair-homotopy type [12; 3.6'] so that, since $\pi(f, g) = \pi(f, Ig) \approx [Jf, g]$, we obtain the following proposition [8; Corollary 2.6].

Proposition 0.2 *If* f *is an h-cofibration or if* g *is an h-fibration then* $\pi(f, g) \approx [f, g]$.

Suppose now that C is an arbitrary category and let P be the associated category of pairs of C. Regarding each object of C as an identity pair enables us to regard C as a full subcategory of P. Moreover C is reflective with associated reflector $f \rightsquigarrow$ codomain f and also coreflective with coreflector $f \rightsquigarrow$ domain f. The embedding is the more interesting because special features of C are often carried over to P. For example it is well-known [17] that if C is abelian then P is abelian. It was clearly the intent of Eckmann and Hilton to extend concepts and, where possible, theorems into the category of pairs. Returning to the case in which C is the pointed homotopy category, Proposition 0.2 implies that the composite $C \to PC \xrightarrow{I} HPC$ is an embedding since identity maps of spaces are cofibrations. Moreover as has been observed in [9; Theorem 7.1] C becomes a reflective (respectively coreflective) subcategory of HPC via the reflector $f \rightsquigarrow$ codomain f (respectively coreflector $f \rightsquigarrow$ domain f). The argument is of course slightly different but the effect is that the program of Eckmann and Hilton still makes sense. In this spirit we examine in §2 the concepts of H-space and H'-space and the interesting phenomenon alluded to in [13] that whereas X is an H-space if and only if the unit of the loop suspension adjunction at X is a coretraction and whereas X is an H'-space if and only if the counit of the loop-suspension adjunction at X is a retraction, the known proofs of these results do not dualize. We show that the relevant extensions of the concepts to HPC already exist but have not been recognized and we obtain corresponding retraction characterizations in terms of the N-P adjunction. We also consider the question of the appropriate definition of Lusternik-Schnirelmann category and cocategory of a map. We show that the method of Ganea [4] for defining cat and cocat via constructing sequences of maps essentially exploits the N-P adjunction.

Eckmann and Hilton [2] succeeded in deriving all the then known exact sequences of homology, cohomology, cohomotopy, etc. as special cases of two dual exact sequences of morphism sets in C and PC. In §1 we show that these can be studied equally well from the point of view of HPC, that in fact some of the relevant proofs become somewhat simpler.

1. Excision morphisms

If Σ denotes the (reduced) suspension functor we shall use the notation $\pi_n(f, g) = \pi(\Sigma^n f, g)$ and $\Pi_n(f, g) = [\Sigma^n f, g]$ to distinguish morphism sets in HPC and PC. For spaces X, Y we have, as remarked in the introduction, $\pi_n(X, Y) = \Pi_n(X, Y)$. If $g : E \to B$ is a pair then Eckmann and Hilton [2] define

$$(1.1) \qquad \Pi_n(X, g) = [\Sigma^{n-1} i, g] ,$$

where $i : X \to CX$ is the pair including X into the cone on X. Since CX is the cofibre of the identity pair X, we recognize that $i = PX$ and we have the following proposition.

Proposition 1.2 $\Pi_n(X, g) \approx \pi_{n-1}(X, F_g)$, *where* F_g *denotes the homotopy fibre of* g.

Proof Let $*X$ denote the inclusion of the base point $*$

into X. Since i is a cofibration, applying Proposition 0.2 we have $\Pi_n(X, g) = \pi_{n-1}(PX, g) \approx \pi_{n-1}(X, Ng) = \pi_{n-1}(P(*X), Ng) \approx \pi_{n-1}(*X, N^2g) = \pi_{n-1}(X, F_g)$. The last equality follows directly from the respective definitions. Alternatively it is a special case of [10; Lemma 1.2 (i)].

Remark 1.3 Proposition 1.2 enables us to recognize the Eckmann-Hilton sequence $S_*(g)$ [12]

$$\to \Pi_n(X, E) \overset{g_*}{\to} \Pi_n(X, B) \overset{J}{\to} \Pi_n(X, g) \overset{\partial}{\to} \Pi_{n-1}(X, B_1) \to$$

as being essentially the dual Puppe sequence

$$\to \pi_n(X, E) \to \pi_n(X, B) \to \pi_{n-1}(X, F_g) \to \pi_{n-1}(X, B) \to .$$

Of course $S_*(g)$ remains convenient as it lends itself so elegantly to specialization. We need not however provide a separate proof of exactness.

Remark 1.4 The reader should be warned that the notation 1.1 is ambiguous. If X is interpreted as the identity pair, which is certainly a cofibration, then $\Pi_n(X, g) \approx \pi_n(X, g) = \pi_n(P(*X), g) \approx \pi_n(*X, Ng) = \pi_n(X, E)$ in conflict with Proposition 1.2. For the remainder of this section we shall use the interpretation 1.1.

Remark 1.5 Proposition 1.2 can be used to derive the formulae [12; 3.1'] :

(1.5.1) *If* $g : * \to B$ *then* $\Pi_n(A, g) \approx \Pi_n(A, B)$.

(1.5.2) *If* $g : B \to *$ *then* $\Pi_n(A, g) \approx \Pi_{n-1}(A, B)$.

Remark 1.6 Let $F = g^{-1}(*)$ and let $\nu : F \to E$ denote the inclusion map. Then the pair maps

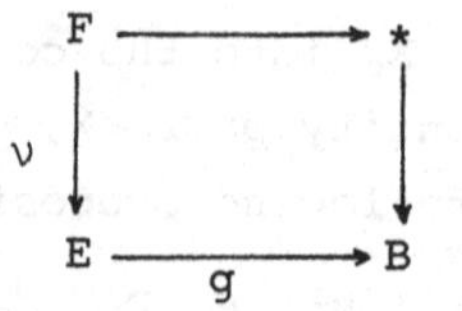

and

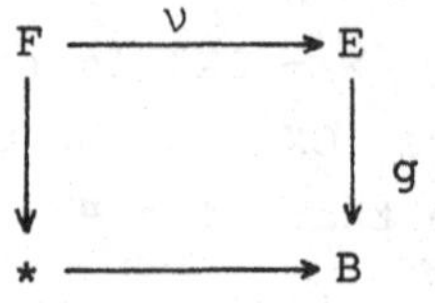

induce the *excision morphisms* $\varepsilon_1^- : \Pi_n(A, \nu) \to \Pi_n(A, B)$ and $\varepsilon_2^- : \Pi_{n-1}(A, F) \to \Pi_n(A, g)$ respectively and we obtain the following short proof of $[12;$ Theorem $3.3']$.

Proposition 1.6.1   *If* g *is a fibre map then* ε_1^- *and* ε_2^- *are isomorphisms.*

Proof   If g is a fibre map then it is well-known $[12;$ Corollary $(3.7)^-$, dual$]$ that ν is homotopy equivalent to Ng. Hence $\Pi_n(A, \nu) \approx \pi_{n-1}(A, F_\nu) \approx \pi_{n-1}(A, F_{Ng}) \approx \pi_{n-1}(A, \Omega B) \approx \pi_n(A, B)$. Similarly ε_2^- is equivalent to $\pi_{n-1}(A, F_g) \to \Pi_n(A, g)$ which is an isomorphism by Proposition 1.2.

2. Cyclic and cocyclic maps

In this section we shall assume that all spaces have the pointed homotopy type of a CW complex.

We recall that a map $f : A \to X$ is *cyclic* if there exists an *associated map* $F : X \times A \to X$ such that the diagram

(2.1)

$$
\begin{array}{ccc}
X \times A & \xrightarrow{\ F\ } & X \\
\big\uparrow{\scriptstyle j} & \nearrow{\scriptstyle \nabla(X \vee f)} & \\
X \vee A & &
\end{array}
$$

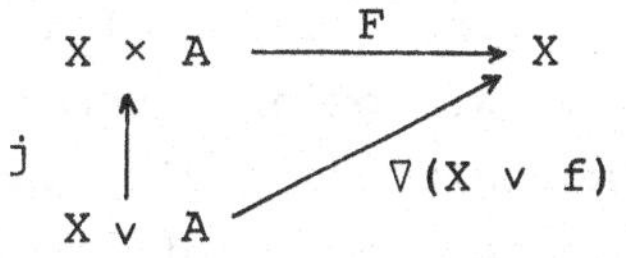

is homotopy commutative, where j is the inclusion and ∇ the codiagonal map. Cyclic maps were first studied by Gottlieb $[6]$ and Varadarajan $[20]$. A convenient survey of the known results can be found in $[15]$. The property of being cyclic is invariant under isomorphism in HPC.   Indeed we have :

Lemma 2.2   *If* $(\phi, \psi, \{h_t\}) : f \to g$ *is a homotopy pair map, if* ϕ *has a left homotopy inverse* μ *and if* g *is cyclic then* f *is cyclic.*

If F' is a map associated to g,   choose $\mu F'(\phi \times \psi)$ as associated map for f.   .
We may clearly associate with $(\phi, \psi, \{h_t\}) : f \to g$ the map $\phi :$ domain (f) $\to$ domain (g).   This rule gives rise to the *domain restriction functor* $d :$ HPC $\to$ HC.
Let $\eta : 1 \to NP$ denote the unit of the NP adjunction. Then we have the following theorem.

Theorem 2.3 *A map* $f : A \to X$ *is cyclic if and only if* $d\eta Pf$ *has a left inverse in* HC.

Remark 2.4 Theorem 2.3 can be regarded as an extension to the pair category of the result that X is an H-space if and only if the unit of the loop suspension adjunction at X is a coretraction, for we have $PX \approx X*$, $P^2X \approx *(\Sigma X)$, $NP^2X \approx (\Omega\Sigma X)*$ and $d\eta PX$ is the unit of the $\Omega\Sigma$ adjunction.

Before proceeding to the proof of Theorem 2.3 it will be convenient to recall some details related to the functors P and N. There is a commutative diagram

$$\begin{array}{ccc} C_f & = & C_f \\ {\scriptstyle P^2f}\downarrow & & \downarrow{\scriptstyle Qf} \\ C_{Pf} & \xrightarrow{Rf} & \Sigma A \end{array} ,$$

where Qf shrinks X to $*$ and Rf shrinks CX to $*$ [18; (5)]. Since P^2 is functorial it follows easily that Q is functorial. Moreover, Rf being a homotopy equivalence, we obtain a natural isomorphism $\sigma : P^2 \to Q$. Dually, $(Nf)^{-1}(*) = \Omega X$ and if we denote by Mf the inclusion of ΩX into F_g we obtain a diagram

$$\begin{array}{ccc} \Omega X & \longrightarrow & F_{Nf} \\ {\scriptstyle Mf}\downarrow & & \downarrow{\scriptstyle N^2f} \\ F_f & = & F_f \end{array} \qquad (2.5)$$

and a natural isomorphism $\tau : M \to N^2$. Then in the diagram

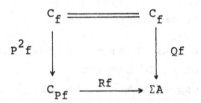

(2.6) 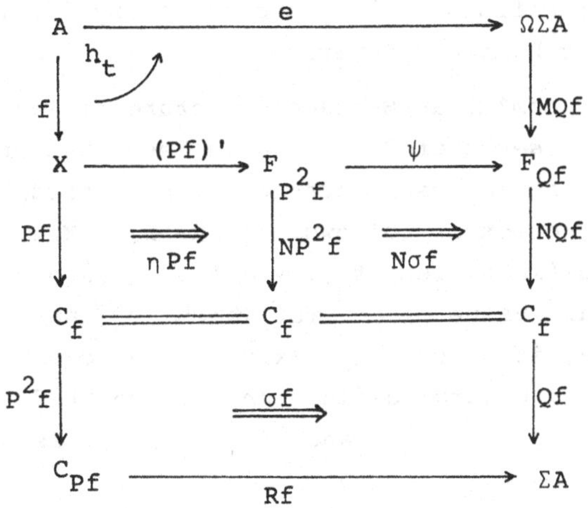

the lower and the two central squares are commutative. The maps $(Pf)'$, ψ and e are given by equations

$$(Pf)'(x) = ((Pf)(x), (x, -)) = (x, (x, -)) \quad (x \in X)$$

$$\psi(z, \lambda) = (z, (Rf)\lambda) \quad (z \in C_f, \lambda \in C_{Pf}{}^I)$$

$$e(a) = (a, -) \quad (a \in A) \quad ,$$

where $(x, -)$ indicates the path $t \mapsto (x, t)$ in C_{Pf}. The homotopy h_t is given by $h_t(a) = ((a, 1-t), a_t)$, where a_t is the path $s \mapsto (a, 1-t + ts)$ in ΣA.

Now Brayton Gray [7; Theorem 4.2] has proved that there exists a homotopy equivalence $\phi : (X, A)_\infty \to F_{Qf}$, where $(X, A)_\infty$ is his relative version of the James construction. Examining his map ϕ it can be checked that $\psi(Pf)' : X \to F_{Qf}$ factors into the inclusion $X \to (X, A)_\infty$ followed by ϕ.

Proof of Theorem 2.3 It is well-known that MQf is cyclic. Suppose that $d\eta Pf$ has a left inverse in HC. Then, since σ is a natural isomorphism, ψ and consequently $\psi(Pf)'$ have left homotopy inverses. Lemma 2.2 now implies that f is cyclic. Conversely suppose that f is cyclic. Without loss of generality we can assume that f is a cofibration (and therefore an inclusion) and that the corresponding diagram 2.1 is strictly commutative. Then A acts on X in the sense of [7; Definition 3.2] and hence, by [7; Theorem 3.2] $(Pf)'$ has a left homotopy inverse.

A space X admits an H-space structure if and only if its cocategory in the sense of Ganea [4] is less than or equal to 2. Ganea's definition can conveniently be expressed in terms of the PN adjunction as follows. Let $f_1 = X^*$, and if f_n is already defined, let $f_{n+1} = d\eta f_n$. Then cocat X is equal to the least value of n for which f_n has a left homotopy inverse, if no such f_n exists then cocat $X = \infty$. Since $X^* \approx PX$, a natural definition for cocat f is obtained by setting $f_1 = Pf$ and $f_{n+1} = d\eta f_n$ as before.

Corollary 2.5 *A map* $f : A \to X$ *is cyclic if and only if* cocat $f \leqslant 2$.

The notion of cyclic map can clearly be dualized as follows. A map $g : X \to A$ is *cocyclic* if and only if there exists an *associated map* $G : X \to X \vee A$ such that the diagram

$$(2.6) \qquad \begin{array}{ccc} X & \xrightarrow{\;G\;} & X \vee A \\ & \searrow^{(X \times g)\Delta} & \downarrow j \\ & & X \times A \end{array}$$

is homotopy commutative. The property of being cocyclic is HPC invariant for we have a dual of lemma 2.2.

Lemma 2.7 *If* g *is cocyclic, if* $(\phi, \psi, \{h_t\}) : g \to f$ *is such that* ψ *has a right homotopy inverse then* f *is cocyclic.*

A dual to Theorem 2.3 can now be stated in terms of the codomain restriction functor $c : HPC \to HC$. Let $\varepsilon : PN \to 1$ denote the counit of the NP adjunction.

Theorem 2.8 *A map* $g : X \to A$ *is cocyclic if and only if* $c\varepsilon Ng$ *has a right inverse in* HC.

Remark 2.9 Since $c\varepsilon NX$ is the counit of the $\Omega\Sigma$ adjunction we recover the result that X is an H'-space if and only if the counit of the loop suspension adjunction at X has a right inverse in HC.

Proof of Theorem 2.8   Suppose that $c\varepsilon Ng$ has a right inverse. It is well known and essentially due to Puppe [18] that Qf is cocyclic for every map f. Let $h : C_{N^2 g} \to X$ be a representative of $c\varepsilon Ng$. Then since $g \cdot h \cdot PN^2 g \simeq 0$ and ΣF_{Ng} is the cofibre of $PN^2 g$ there exists a map $h' : \Sigma F_{Ng} \to A$ such that the lower rectangle of the following diagram is homotopy commutative.

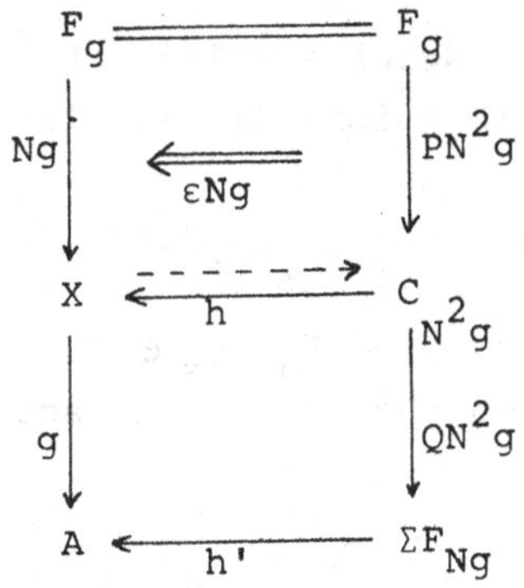

Applying Lemma 2.7 we have that g is cocyclic. Conversely suppose that g is cocyclic and consider the standard homotopy pullback diagram :

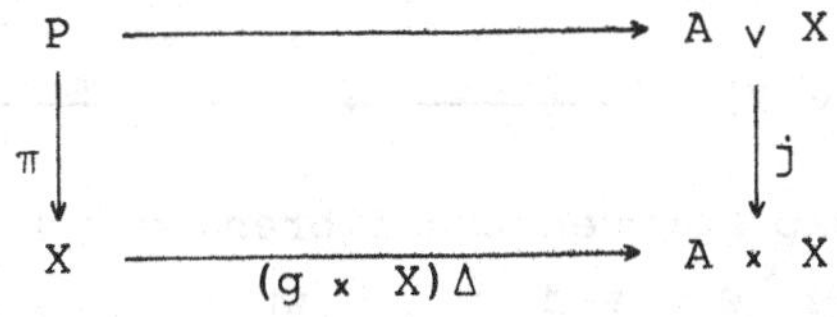

An associated map G provides a homotopy diagonal for the above rectangle and therefore a homotopy section for π exists. To complete the proof it is sufficient to show that π is isomorphic in HPC to h. The following argument modifies and specializes the construction of Gilbert [5; Proposition 3.3]. P is the total space obtained by converting j into a fibration and pulling back over $(g \times X)\Delta$. Thus P is the relative path space $E(A \times X, \Delta g, A \vee X)$, where $\Delta g \subset A \times X$ is the image of $(g \times X)\Delta$, and $\pi(\xi) = \pi_2\xi(0)$, where $\pi_2 : A \times X \to X$ is the projection. In view of the natural isomorphism $\tau : M \to N^2$, we have $h \approx k$, where $k : C_{Mg} = F_g \cup C\Omega A \to X$ agrees with Ng and maps $C\Omega A$ to $*$. To compare k and π, convert k into the fibre map $v : U \to X$, where

$$U = \{(s\mu, \nu) \in F_g \cup C\Omega A \subset CF_g \times X^I \mid k(s\mu) = \nu(1)\}$$

and $v(s\mu, \nu) = \nu(0)$. Then v has fibre

$$V = \{(s\mu, \nu) \in F_g \cup C\Omega A \times X^I \mid k(s\mu) = \nu(1),\ \nu(0) = *\}.$$

Let $\lambda : \{(\mu, \nu) \in F_g \times X^I \mid Ng(\mu) = \nu(1)\} \to F_g{}^I$ be a path lifting function for the fibration Ng and let $w : U \to P$ be given by

$$w(s\mu, \nu) = (\lambda(\mu, -\nu)(1)_s, \nu),$$

where for any point $\xi = (x, \rho) \in F_g$, $\xi_s \in A^I$ is the path such that $\xi_s(t) = \rho(st)$. Then the right hand square in the following diagram commutes.

$$\begin{array}{ccccc}
V & \longrightarrow & U & \xrightarrow{\ v\ } & X \\
\downarrow{\scriptstyle w'} & & & & \\
\Omega A \times E(X, *, X) \cup C\Omega A \times \Omega X & & \downarrow{\scriptstyle w} & & \| \\
\downarrow{\scriptstyle w''} & & & & \\
E(A \times X, *, A \vee X) & \longrightarrow & P & \xrightarrow{\ \pi\ } & X
\end{array}$$

Hence w induces a map between the fibres, which can be factored into two maps w' and w'', where

$$w'(s\mu,\nu) = (s\lambda(\mu, -\nu)(1), \nu)$$

$$w''(s\xi, \nu) = (\xi_s, \nu).$$

By the arguments used in the proof of [4; Theorem 1.1], both w' and w'' are weak homotopy equivalences. An application of the 5-lemma shows that w is a weak homotopy equivalence. Since C_{N^2g} and P have the homotopy type of CW-complexes it follows that $h \approx k \approx v \approx \pi$ in HPC, completing the proof of Theorem 2.8.

Dual considerations and also the force of Ganea's

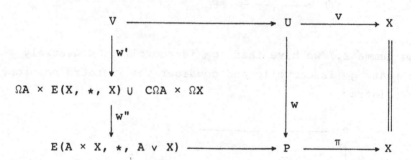

initiative now suggest the following definition of Lusternik Schnirelmann category of a map g. Let $g_1 = Ng$, and if g_n is already defined let $g_{n+1} = c\varepsilon g_n$ Set G cat g = n if n is the least value of m for which g_m has a right inverse in HC, otherwise set G cat g = ∞.

Corollary 2.10 *A map* g *is cocyclic if and only if* G cat g ⩽ 2.

In view of the above it is tempting to define cat g = G cat g. This cannot be done however as cat g already has an established meaning [1]. To see that the two do not coincide, consider a sphere bundle $p : E = S^m \cup e^n \cup e^{m+n} \to S^n$ with the property that $p^*\iota = \alpha \in H^n(E)$, where α, β, γ are generators in cohomology with $\gamma = \alpha \cup \beta$ [14]. Since cat $(S^n) = 2$, by a standard result we have cat p ⩽ 2. However p is not a cocyclic map since any associated map $E \to E \vee S^n$, if such existed, would fail to preserve cup products. It follows that G cat p ⩾ 3.

To enable more effective comparison of G cat g and cat g, one may give a "Whitehead type" characterization of G cat g. Let $\Pi(n)$ denote the product of one copy of A and n-1 copies of X, let $g_\Delta : X \to \Pi(n)$ be the map which followed by projection on to A yields g and which followed by projection onto each factor X yields the identity map. Let $j : T \to \Pi(n)$ denote the inclusion of the fat wedge. Then we have the following.

Theorem 2.11 G cat g ⩽ n *if and only if* g_Δ *factors through* $j : T \to \Pi(n)$ *up to homotopy.*

Proof Convert j into a fibration and let $\pi : P \to X$ denote its pull-back over g_Δ. Convert also g_n into a fibration. Then it is claimed that these fibrations are homotopy equivalent. Indeed we need only complete the modification of Gilbert's inductive argument in the proof of [5; Proposition 3.3], of which the first stage has already been given in the proof of Theorem 2.8.

An obvious consequence of Theorem 2.11 is the following.

Corollary 2.12 *If* $g : X \to A$ *then* $\mathrm{cat}\, g \leq G\,\mathrm{cat}\, g \leq \mathrm{cat}\, X$.

If $g : X \to A$ is homotopically trivial it is easy to see that $G\,\mathrm{cat}\, g = 1$, hence we can have $G\,\mathrm{cat}\, g < \mathrm{cat}\, A$. This example also shows that g cocyclic does not imply that A is an H'-space.

3. Fibrations that are also cofibrations

Let $F \overset{i}{\to} E \overset{p}{\to} B$ be a fibration. Then it is *also a cofibration* in the sense of Milgram [16] if there exists a homotopy commutative diagram

(3.1)
$$\begin{array}{ccccc} F & \xrightarrow{\ i\ } & E & \xrightarrow{\ Pi\ } & C_i \\ \Big\| & h_t \curvearrowright & \Big\downarrow E & & \Big\downarrow k \\ F & \xrightarrow{\ i\ } & E & \xrightarrow{\ p\ } & B \end{array}$$

in which the map k is a homotopy equivalence. Examples have been given by Milgram loc.cit., Hausmann and Husemoller [11], Schiffman [19] and the phenomenon studied in some depth by Wojtkowiak [21].

Proposition 3.2 *The homotopy class of* k *in diagram* 3.1 *is equivalent in* HC *to* εp.

Proof First observe that the homotopy commutative left hand square of 3.1 defines an automorphism of i in HPC. Hence replacing the left hand square of 3.1 by a strictly commutative square affects k only by composing it with a homotopy equivalence. Next note that i is isomorphic in HPC to Np. Hence replacing the right hand square of 3.1 by $\varepsilon p : PNp \to p$, replaces k by an equivalent class. The following is an immediate consequence.

Proposition 3.3 *A fibration* p *is also a cofibration if and only if the counit of the* NP *adjunction at* p *is an isomorphism.*

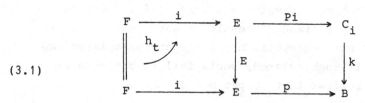

Since any map is isomorphic in HPC to a fibration, the fibrations that are also cofibrations are completely determined by the subcategory of HPC orthogonal to ε.

Proposition 3.3 enables some elucidation of the dual phenomenon : *a cofibration* j *is also a fibration* if the unit of the NP adjunction at j is an isomorphism. It is well known that adjoint functors, restricted to the respective full subcategories of objects fixed by the unit and counit, yield category equivalences. It follows that every fibration that is also a cofibration can be obtained by applying the functor P to a cofibration that is also a fibration and then converting to a fibration. Conversely every cofibration that is also a fibration can be obtained by applying the functor N to a fibration that is also a cofibration and then converting to a cofibration. While these observations are somewhat trivial, at least they bring out the point that this is one of the happy instances in which the Eckmann-Hilton duality is categorical.

References

[1] BERSTEIN, I. and GANEA, T. The category of a map and of a cohomology class. *Fundamenta Math.* 50 (1961/2), 265-279.

[2] ECKMANN, B. and HILTON, P.J. Groupes d'homotopie et dualité. *C.R. Acad. Sci. (Paris)* 246 (1958), 2444, 2555, 2991.

[3] GABRIEL, P. and ZISMAN, M. Calculus of fractions and homotopy theory. *Ergebnisse der Math. und ihre Grenzgebiete* 35 (1967). Springer Verlag.

[4] GANEA, T. A generalization of the homology and homotopy suspension. *Comment. Math. Helv.* 39 (1965), 295-322.

[5] GILBERT, W.J. Some examples for weak category and conilpotency. *Illinois. J. M.* 12 (1968), 421-432.

[6] GOTTLIEB, D.H. A certain subgroup of the fundamental group. *Amer. J. Math.* 87 (1965), 840-856

[7] GRAY, B. On the homotopy groups of mapping cones. *Proc. London Math. Soc.* (3) 26 (1973), 497-520.

[8] HARDIE, K.A. On the category of homotopy pairs. *Topology and its applications* 14 (1982), 59-69.

[9] HARDIE, K.A. and JANSEN, A.V. The Puppe and Nomura operators in the category of homotopy pairs. Proc. Conf. Categorical Aspects of Topology and Analysis, Carleton (1980). *Lecture Notes in Math.* 915 Springer Verlag, 112-126.

[10] HARDIE, K.A. and JANSEN, A.V. Toda brackets and the category of homotopy pairs. Proc. Symp. Categorical Algebra and Topology, Cape Town (1981). *Quaestiones Math.* 6 (1983), 107-128.

[11] HAUSMANN, J.C. and HUSEMOLLER, D. Acyclic maps. *Enseignment Math.* (2) 25 (1979), 53-75.

[12] HILTON, P. Homotopy theory and duality. Gordon and Breach Science Publishers, Inc. (1965).

[13] HILTON, P. Duality in homotopy theory : a retrospective essay. *J. Pure and Applied Algebra* 19 (1980), 159-169.

[14] JAMES, I.M. Note on cup-products. *Proc. Amer. Math. Soc.* 8 (1957), 374-383.

[15] LIM, K.L. On cyclic maps. *J. Aust. M. Soc.* 32 (1982), 349-357.

[16] MILGRAM, R.J. Surgery with coefficients. *Ann. of Math.* (1974), 194-248.

[17] PRESSMAN, I.S. Functors whose domain is a category of morphisms. *Acta Math.* 118 (1967), 223-249.

[18] PUPPE, D. Homotopiemengen und ihre induzierten Abbildungen I. *Math. Zeitschr.* 69 (1958), 299-344.

[19] SCHIFFMAN, S.J. A mod p Whitehead theorem. *Proc. American Math. Soc.* 82 (1981), 139-144.

[20] VARADARAJAN, K. Generalized Gottlieb groups. *J. Indian Math. Soc.* 33 (1969), 141-164.

[21] WOJTKOWIAK, Z. On fibrations which are also cofibrations. *Quart. J. Math. Oxford* (2) 30 (1979), 505-512.

Grants to the Topology Research Group from the University of Cape Town and the South African Council for Scientific and Industrial Research are acknowledged.

The Department of Mathematics
University of Cape Town
Rondebosch 7700
Republic of South Africa.

PROFINITE CHERN CLASSES FOR GROUP REPRESENTATIONS

BENO ECKMANN and GUIDO MISLIN

Eidgenössische Technische Hochschule Zürich

To our friend Peter Hilton, on his sixtieth birthday

1. INTRODUCTION

1.1. Let G be a (discrete) group and $\varrho : G \to GL_m(\mathbb{C})$ a complex representation of G of degree m. With ϱ one associates Chern classes $c_j(\varrho) \in H^{2j}(G;Z)$ in the cohomology of G, as follows: The map $B\varrho : BG \to BGL_m(\mathbb{C})$ of classifying spaces induces a $\mathbb{C}^m$-bundle over $BG = K(G,1)$ (Eilenberg-MacLane complex for G), and the $c_j(\varrho)$ are the Chern classes of that vector bundle, $j = 1,2,3,\ldots$. The bundle being flat there are many groups for which the Chern classes have finite order; indeed it is well-known that the image of $c_j(\varrho)$ in $H^{2j}(G;\mathbb{Q})$ under the inclusion coefficient map $Z \to \mathbb{Q}$ is 0, which implies, under suitable finiteness conditions on G, that $c_j(\varrho)$ is a torsion element.

This is trivially so if G is a finite group. In that case it was shown in [5] that if the representation ϱ is realizable over a number field $K \subset \mathbb{C}$, then there exists a precise bound for the order of $c_j(\varrho)$ depending on K and j only: An integer $\bar{E}_K(j)$ is described in [5] with the property that $\bar{E}_K(j)\, c_j(\varrho) = 0$ for all finite groups G and all K-representations ϱ of G ; and that $\bar{E}_K(j)$, or $\frac{1}{2}\bar{E}_K(j)$ depending on properties of the number field K , is best possible in that sense. Namely, let $e_K(n)$ for a positive integer n denote the exponent of $Gal(K(\zeta_n)/K)$, the Galois group over K of the n-th cyclotomic extension $K(\zeta_n)$ where ζ_n is a primitive n-th root of unity; then

$$\bar{E}_K(j) = \max\{n \text{ with } e_K(n) \text{ dividing } j\} .$$

In [5] the prime factorization of $\bar{E}_K(j)$ has been expressed in terms of numerical invariants attached to K . In particular, for $K = \mathbb{Q}$ and j even one has $\bar{E}_{\mathbb{Q}}(j)$ = denominator of $\frac{B_j}{2j}$ where B_j is the j-th Bernoulli number ($B_2 = \frac{1}{6}$, $B_4 = \frac{1}{30}$, $B_6 = \frac{1}{42}$ etc.) Some number-theoretic properties of $\bar{E}_K(j)$ have been investigated in Section 6 of [5]. We note that the numbers $\bar{E}_K(j)$ are equal to the $w_j(K)$ in Cassou-Noguès [3] (see also [7]).

<u>1.2</u>. The purpose of the present paper is to discuss corresponding bounds for <u>arbitrary</u>, finite or infinite, groups and their finite-dimensional complex representations. Here a result of the same generality for the order of the $c_j(\varrho) \in H^{2j}(G;\mathbb{Z})$ is not known. However, it will turn out that corresponding precise bounds can be established for the order of the <u>profinite</u> Chern classes $\hat{c}_j(\varrho) \in H^{2j}(G;\hat{\mathbb{Z}})$. As usual $\hat{\mathbb{Z}}$ denotes the ring of profinite integers, $\hat{\mathbb{Z}} = \varprojlim \mathbb{Z}/n\mathbb{Z}$. The Chern class $\hat{c}_j(\varrho)$ is the image of $c_j(\varrho)$ under the obvious coefficient homomorphism $\mathbb{Z} \to \hat{\mathbb{Z}}$. One again considers a number field K and a representation $\varrho : G \to GL_m(\mathbb{C})$. To

formulate the result we do not assume ϱ to be realizable over K but only to have its character values in K ; this is natural in view of the fact (see Section 2) that the representation ring $R(G)$ is isomorphic to the character ring of G , and that the Chern classes only depend on the element of $R(G)$ corresponding to ϱ . Under that character assumption one obtains precise bounds for the order of the $\hat{c}_j(\varrho)$, and they are again the integers $\bar{E}_K(j)$ above:

Main Theorem. Let $K \subset \mathbb{C}$ be a number field.

A) For any group G and any representation $\varrho : G \to GL_m(\mathbb{C})$ of arbitrary degree m whose character takes its values in K one has

$$\bar{E}_K(j)\hat{c}_j(\varrho) = 0 \ , \quad j = 1,2,3,\ldots \ .$$

B) The bound $\bar{E}_K(j)$ is the best possible fulfilling A).

1.3. Remarks. 1) The problem of the best possible bound for the $\hat{c}_j(\varrho)$ of representations realizable over K is left open. Clearly that bound is $\bar{E}_K(j)$ or perhaps, depending on the field, $\frac{1}{2}\bar{E}_K(j)$. Cases of number fields K where it must be $\bar{E}_K(j)$ can be deduced from the explicit discussion in [5].

2) The kernel of the coefficient homomorphism $H^{2j}(G;Z) \to H^{2j}(G;\hat{Z})$ consists of all elements of $H^{2j}(G;Z)$ which are infinitely divisible. Thus if there are no such elements in $H^{2j}(G;Z)$, the Chern classes $c_j(\varrho)$ and $\hat{c}_j(\varrho)$ have the same order so that we get information about the $c_j(\varrho)$ themselves. This is the case, for example, if the

group G is virtually of type (FP) ; but, of course, that condition is too strong. In particular, cf. the appendix concerning $c_1(\varrho)$.

3) The arithmetic group $GL_n(O_K)$, where O_K is the ring of integers of the number field K , is virtually of type (FP) (see [8]). Thus for any representation ϱ of $GL_n(O_K)$ the order of $c_j(\varrho)$ is finite and divides $\bar{E}_K(j)$. In particular, if ι denotes the canonical representation $GL_n(O_K) \to GL_n(\mathbb{C})$ for large n , the order of the Chern class $c_j(\iota)$ is equal to $\bar{E}_K(j)$ unless j is even and K formally real, where it is either $\bar{E}_K(j)$ or $\frac{1}{2}\bar{E}_K(j)$ (cf. [5], Section 5; the precise general answer is not known).

1.4. The method for obtaining bounds for the order of Chern classes of group representations is based on "Galois invariance". One applies an automorphism σ of $\mathbb{C}$ to the representation ϱ ; if the character takes its values in K and if σ fixes the elements of K , i.e., $\sigma \in Gal(\mathbb{C}/K)$ then ϱ and its image under σ belong to the same element of $R(G)$ and thus have the same Chern classes. On the other hand, the effect of σ on the (profinite) Chern classes can be described explicitely. This method goes back to Grothendieck [6] where it was applied to "algebraic p-adic Chern classes" and then carried over to ordinary Chern classes by means of a comparison theorem for torsion elements.

We proceed differently. In the case of finite groups a direct approach has been used in [5]. It consists in identifying the Galois action with certain Adams operations on the representation ring and on the corresponding flat bundles; then one has well-known properties of Chern classes under these operations. In the general case of arbitrary groups and profinite Chern classes, the effect of $\sigma \in Gal(\mathbb{C}/K)$ on $\hat{c}_j(\varrho)$ is multiplication by a unit $\hat{k}(\sigma)^j \in \hat{Z}$ determined by the action of σ on the roots of unity. This is shown by applying a construction of Sullivan's [10] to representations of groups G of geometrically finite type (i.e., admitting a finite Eilenberg-MacLane complex $K(G,1)$). Then, we show that for our purpose an arbitrary group can be approximated by groups of geometrically finite type.

2. THE CHARACTER RING

2.1. A representation $\varrho : G \to GL_m(\mathbb{C})$ defines a G-action on $\mathbb{C}^m$, turning $\mathbb{C}^m$ into a $\mathbb{C}G$-module $V = V(\varrho)$; in the following all $\mathbb{C}G$-modules are meant to be of finite dimension over $\mathbb{C}$. The complex representation ring $R(G)$ of the group G is defined to be the ring additively generated by all $\mathbb{C}G$-modules, with relations $[V] = [V'] + [V'']$ for every short exact sequence $0 \to V' \to V \to V'' \to 0$ of $\mathbb{C}G$-modules where $[V]$ denotes the image of V in $R(G)$. The multiplication in $R(G)$ is defined by the tensor product $V \otimes W$ over $\mathbb{C}$ of $\mathbb{C}G$-modules V and W.

Given $V = V(\varrho)$ one considers a composition series

$$V_o = 0 \subset V_1 \subset V_2 \subset \ldots \subset V_{n-1} \subset V_n = V .$$

Then $[V] = \sum_{\nu=1}^{n} [V_\nu/V_{\nu-1}]$ where all $V_\nu/V_{\nu-1}$ are simple $\mathbb{C}G$-modules, i.e., isomorphic to $V(\varrho_\nu)$ for an irreducible representation ϱ_ν ; and $[V] = [\bigoplus_{\nu=1}^{n} V(\varrho_\nu)]$. Thus from the point of view of $R(G)$ all $\mathbb{C}G$-modules are semi-simple; in other words, we can restrict attention to completely reducible representations. By the Jordan-Hölder theorem the ϱ_ν are uniquely determined up to equivalence and order. The elements $[V_\alpha]$ where the V_α are all inequivalent simple $\mathbb{C}G$-modules form an additive basis of $R(G)$.

2.2. The character χ_ϱ of the representation $\varrho : G \to GL_m(\mathbb{C})$ only depends on $V = V(\varrho)$; we also write χ_V for χ_ϱ . If $0 \to V' \to V \to V'' \to 0$ is a short exact sequence of $\mathbb{C}G$-modules then $\chi_V = \chi_{V'} + \chi_{V''}$. Therefore the passage $V \mapsto \chi_V$ defines an additive homomorphism χ of $R(G)$ into the ring of complex-valued functions on G . Since $\chi_{V \otimes W} = \chi_V \cdot \chi_W$ this is actually a ring homomorphism; its image $\chi(R(G))$ is denoted by $R_\chi(G)$ and called the character ring of G .

Theorem 1. *The map* $\chi : R(G) \to R_\chi(G)$ *is a ring isomorphism*.

Proof. If two completely reducible representations $\varrho_1, \varrho_2 : G \to GL_m(\mathbb{C})$ have the same character $\chi_{\varrho_1} = \chi_{\varrho_2}$ then the $\mathbb{C}G$-modules $V(\varrho_1)$ and $V(\varrho_2)$ are isomorphic (this is a consequence of the double centralizer theorem, cf. [2] chap.VIII, §12, Prop. 3). An arbitrary element $x \in R(G)$ can be written as $x = \Sigma[V_\nu] - \Sigma[W_\nu]$ where all V_ν and W_μ are simple

$\mathbb{C}G$-modules. Then $\chi(x) = \chi([\bigoplus_\nu V_\nu]) - \chi([\bigoplus_\mu W_\mu])$. If we assume $\chi(x) = 0$ then $\bigoplus_\nu V_\nu$ and $\bigoplus_\mu W_\mu$ have the same character and hence are isomorphic by the above remark. It follows that $x = 0 \in R(G)$ and thus χ is injective; it is surjective by definition.

2.3. We now consider the action of an automorphism σ of $\mathbb{C}$ on the representations and the corresponding elements of $R(G)$. The automorphism σ induces a group automorphism σ of $GL_m(\mathbb{C})$. Given a representation $\varrho : G \to GL_m(\mathbb{C})$ we write ϱ^σ for the composite representation $\sigma\varrho$. If we assume that σ leaves fixed all values of the character χ_ϱ then ϱ and ϱ^σ have the same character. By Theorem 1 this implies that $[V(\varrho)] = [V(\varrho^\sigma)] \in R(G)$:

Corollary 2. *Let* $\varrho : G \to GL_m(\mathbb{C})$ *be a representation and* σ *an automorphism of* $\mathbb{C}$ *which fixes all values of* χ_ϱ . *Then* $[V(\varrho)] = [V(\varrho^\sigma)]$.

Remark. In the case of a finite group G it is well-known that ϱ and ϱ^σ in Corollary 2 are actually equivalent. This need, however, not be the case for infinite groups. For example, if $\varrho : \mathbb{Z} \to GL_4(\mathbb{C})$ is given by

$$\varrho(1) = \left(\begin{array}{cc|cc} i & 1 & & \\ 0 & i & & \\ \hline & & -i & 0 \\ & & 0 & -i \end{array}\right)$$

then complex conjugation σ fixes the character values, but ϱ and ϱ^σ are not equivalent.

2.4. The Chern classes of ϱ (cf. Section 1.1) $c_j(\varrho) \in H^{2j}(G;Z)$ depend on $V(\varrho)$ only; we may also write $c_j(V)$ for any $\mathbb{C}G$-module V (of finite dimension over $\mathbb{C}$). Let $c(V) = 1 + c_1(V) + c_2(V) + \ldots \in H^*(G;Z)$ be the total Chern class. If $0 \to V' \to V \to V'' \to 0$ is a short exact sequence then $c(V) = c(V') \cdot c(V'')$ in the cohomology ring $H^*(G;Z)$; this is due to the fact that every short exact sequence of vector bundles over a CW-complex splits as a direct sum. Thus the total Chern class defines a homomorphism of the additive group of $R(G)$ into the multiplicative group of units of the graded ring $H^*(G;Z)$. If two representations ϱ_1, ϱ_2 of G have the same character then the corresponding elements of $R(G)$ coincide, and hence the Chern classes $c_j(\varrho_1)$ and $c_j(\varrho_2)$ are the same, $j = 1,2,3,\ldots$:

Theorem 3. If two representations of the group G have the same character then their Chern classes coincide. In particular, if an automorphism σ of $\mathbb{C}$ fixes the character values of the representation ϱ then ϱ^σ and ϱ have the same Chern classes.

3. GALOIS ACTION AND ROOTS OF UNITY

3.1. Let $\mu(\mathbb{C})$ denote the group of all roots of unity in $\mathbb{C}$. For a number field $K \subset \mathbb{C}$ the following integers have been considered by Soulé [9]:

$$w_j(K) = \text{card}\{z \in \mu(\mathbb{C}) \text{ with } \sigma^j(z) = z \text{ for all } \sigma \in \text{Gal}(\mathbb{C}/K)\}$$

where as usual $\text{Gal}(\mathbb{C}/K)$ denotes the group of automorphisms of $\mathbb{C}$ leaving fixed all elements of K. We write $\mu_n \subset \mu(\mathbb{C})$

for the group of n-th roots of unity in $\mathbb{C}$; if ζ_n is a primitive n-th root of unity, $\mu_n \subset K(\zeta_n)$. The Galois group $Gal(\mathbb{C}/K)$ acts on μ_n by an operation $Gal(\mathbb{C}/K) \to Aut\, \mu_n$, which factors through the surjective restriction map $Gal(\mathbb{C}/K) \to Gal(K(\zeta_n)/K)$ whose action on μ_n is, of course, faithful. Therefore $\sigma^j(z) = z$ for all $z \in \mu_n$ and all $\sigma \in Gal(\mathbb{C}/K)$ if and only if j is a multiple of $e_K(n)$, the exponent of $Gal(K(\zeta_n)/K)$.

The elements of $\mu(\mathbb{C})$ fixed under σ^j , for a given j and all $\sigma \in Gal(\mathbb{C}/K)$, form a finite group which must be cyclic, i.e., equal to μ_n for some n . It follows that the number of elements of $\mu(\mathbb{C})$ fixed under σ^j for all $\sigma \in Gal(\mathbb{C}/K)$ is the greatest integer n such that $e_K(n)$ divides j ; i.e., it is the integer $\bar{E}_K(j)$ defined in the introduction:

Proposition 4. Let K be a number field. The elements of $\mu(\mathbb{C})$ fixed under σ^j for all $\sigma \in Gal(\mathbb{C}/K)$ form a cyclic group of order $\bar{E}_K(j) = w_j(K)$, $j = 1,2,3,\ldots$.

3.2. We now consider the profinite integers $\hat{Z} = \varprojlim_n Z/nZ$; here and throughout this section $\varprojlim_n$ refers to an inverse system indexed by the natural numbers with their divisibility relation.

An automorphism σ of $\mathbb{C}$ acts on $\hat{Z}$, as follows. The action of σ on μ_n is by the k-power map for a certain $k = k_n(\sigma) \in (Z/nZ)^*$. The sequence $k_n(\sigma)$, $n = 1,2,3,\ldots$ defines a unit $\hat{k}(\sigma) \in \hat{Z}^*$ in the ring $\hat{Z}$. If we put $\sigma_*: \hat{Z} \to \hat{Z}$

to be multiplication by $\hat{k}(\sigma)$, this is an automorphism of the additive group of $\hat{Z}$; it is the same as the inverse limit of the automorphism of Z/nZ given by multiplication with $k_n(\sigma)$. Note that one has $\hat{k}(\sigma)^j = \hat{k}(\sigma^j)$, $j = 1,2,\ldots$.

Any inverse limit $A = \varprojlim_n A_n$ of Z/nZ-modules is a $\hat{Z}$-module in the obvious way. Thus an action σ_* of an automorphism σ of $\mathbb{C}$ on A can be defined as above by means of the unit $\hat{k}(\sigma)$ in $\hat{Z}$.

Proposition 5. Let $A = \varprojlim_n A_n$ be an inverse limit of Z/nZ-modules, and $K \subset \mathbb{C}$ a number field. If, for a fixed $j = 1,2,3,\ldots$, an element $a \in A$ fulfills $\hat{k}(\sigma)^j a = a$ for all $\sigma \in \mathrm{Gal}(\mathbb{C}/K)$ then a is of finite order dividing $\bar{E}_K(j)$.

Proof. Put $a = \varprojlim_n a_n$, $a_n \in A_n$. The assumption $\hat{k}(\sigma)^j a = a$ means that $k_n(\sigma)^j a_n = a_n$ for each n and all $\sigma \in \mathrm{Gal}(\mathbb{C}/K)$. If we choose an isomorphism of the cyclic subgroup of A_n generated by a_n onto a subgroup of $\mu(\mathbb{C})$, a_n corresponds to an element $z \in \mu(\mathbb{C})$ fulfilling $\sigma^j z = z$ for all $\sigma \in \mathrm{Gal}(\mathbb{C}/K)$. By Prop. 4 the order of z , and hence of a_n , divides $\bar{E}_K(j)$ and so does the order of $a = \varprojlim_n a_n$.

3.3. The case we are especially interested in is that of cohomology groups $H^i(X;\hat{Z})$ of a CW-complex X with $\hat{Z}$-coefficients. They are $\hat{Z}$-modules, the operation being induced by the multiplication of the coefficients; thus they admit an action of any automorphism σ of $\mathbb{C}$, through multiplying the coefficients $\hat{Z}$ with $\hat{k}(\sigma)$.

It is known that the canonical map

$$H^i(X;\hat{Z}) \to \varprojlim_n H^i(X;Z/nZ)$$

defined by the maps $\hat{Z} \to Z/nZ$ for all n is an isomorphism (Sullivan [10]). Thus $H^i(X;\hat{Z})$ is an inverse limit $\varprojlim_n$ of Z/nZ-modules; the action of $\hat{k} \in \hat{Z}^*$ on $H^i(X;\hat{Z})$ is therefore the inverse limit of actions on these Z/nZ-modules exactly as in 3.2 above.

3.4. We further recall the following result of Sullivan [10] concerning the action of an automorphism σ of $\mathbb{C}$ on the profinite completion of $BGL(\mathbb{C})$, the classifying space of the infinite complex linear group $GL(\mathbb{C})$. The automorphism σ acts on the etale homotopy type of any complex variety defined over $\mathbb{Q}$, in particular of the complex Grassmann manifolds. This action may be used to define an induced action on the profinite completion of the classical homotopy type of the variety. By a limit argument one obtains an action $\sigma : BGL(\mathbb{C})^\wedge \to BGL(\mathbb{C})^\wedge$ where $^\wedge$ denotes profinite completion. Sullivan's result concerns its effect on (even-dimensional) cohomology with $\hat{Z}$-coefficients.

Proposition 6. *The action of the automorphism σ of $\mathbb{C}$ on $BGL(\mathbb{C})^\wedge$ induces in $H^{2j}(BGL(\mathbb{C})^\wedge;\hat{Z})$ the coefficient homomorphism given by multiplication of $\hat{Z}$ with $\hat{k}(\sigma)^j$,* $j = 1,2,3,\ldots$.

4. GEOMETRICALLY FINITE GROUPS

4.1. A group G is said to be of geometrically finite type if it admits a finite Eilenberg-MacLane complex $K(G,1)$; this is the case if and only if G is of type (FF) (i.e., admits a finite free resolution of Z over ZG) and finitely presentable.

Let G be such a group, and $\varrho : G \to GL_m(\mathbb{C})$ a representation, σ an automorphism of $\mathbb{C}$. We consider the maps of classifying spaces $B\varrho : K(G,1) \to BGL_m(\mathbb{C})$ and $B\varrho^{\sigma} : K(G,1) \to BGL_m(\mathbb{C})$, and their profinite completions $(B\varrho)^{\wedge}$ and $(B\varrho^{\sigma})^{\wedge}$; moreover, the canonical map $BGL_m(\mathbb{C})^{\wedge} \to BGL(\mathbb{C})^{\wedge}$, and the self-map σ of $BGL(\mathbb{C})^{\wedge}$, see Section 3.4. Using techniques of etale homotopy (similar to those used in [4]; we will come back to this argument in a separate paper), and the fact that $K(G,1)$ is compact, it follows that the two maps

$$K(G,1)^{\wedge} \xrightarrow{(B\varrho)^{\wedge}} BGL_m(\mathbb{C})^{\wedge} \longrightarrow BGL(\mathbb{C})^{\wedge} \xrightarrow{\sigma} BGL(\mathbb{C})^{\wedge}$$

and

$$K(G,1)^{\wedge} \xrightarrow{(B\varrho^{\sigma})^{\wedge}} BGL_m(\mathbb{C})^{\wedge} \longrightarrow BGL(\mathbb{C})^{\wedge}$$

are homotopic; thus they induce the same homomorphism in cohomology with $\hat{Z}$-coefficients. The second map yields the profinite Chern classes $c_j(\varrho^{\sigma}) \in H^{2j}(G;\hat{Z})$, while the first map yields the Chern classes $\hat{c}_j(\varrho)$ multiplied by $\hat{k}(\sigma)^j$, according to Proposition 6:

Theorem 7. Let G be a group of geometrically finite type, $\varrho : G \to GL_m(\mathbb{C})$ a representation, and σ an automorphism of $\mathbb{C}$. Then the profinite Chern classes fulfill

$$\hat{c}_j(\varrho^\sigma) = \hat{k}(\sigma)^j \hat{c}_j(\varrho) \in H^{2j}(G;\hat{Z}) .$$

4.2. In order to carry over the statement of Theorem 7 to arbitrary groups we use the following approximation method.

Let G be a group, and $X = K(G,1)$ an Eilenberg-MacLane complex for G . We may consider X as a union $\cup_\alpha X_\alpha$ of connected finite subcomplexes X_α . By the construction of Baumslag-Dyer-Heller [1] there is, for each α , a geometrically finite group G_α and an acyclic map $g_\alpha : K(G_\alpha,1) \to X_\alpha$; the g_α induce isomorphisms $g_\alpha^* : H^i(X_\alpha;\hat{Z}) \to H^i(G_\alpha;\hat{Z})$ for all i . Let $f_\alpha : G_\alpha \to G$ be the homomorphism of fundamental groups induced by $\iota_\alpha g_\alpha : K(G_\alpha,1) \to X_\alpha \to X$ where ι_α is the inclusion map $X_\alpha \to X$.

The cohomology maps ι_α^* define a homomorphism $H^i(X;\hat{Z}) \to \varprojlim_\alpha H^i(X_\alpha;\hat{Z})$ which is an isomorphism (Sullivan [10]; one uses the fact that $H^i(X_\alpha;\hat{Z})$ can be given a compact topology). It follows that the canonical map $H^i(X;\hat{Z}) \to \varprojlim_\alpha H^i(X_\alpha;\hat{Z}) \to \prod_\alpha H^i(X_\alpha;\hat{Z})$ is injective. In other words, the family of homomorphisms $f_\alpha : G_\alpha \to G$ defines an injective homomorphism $\{f_\alpha^*\} : H^i(G;\hat{Z}) \to \prod_\alpha H^i(G_\alpha;\hat{Z})$, for all i :

Theorem 8. Let G be an arbitrary group. There exists a family of homomorphism $f_\alpha : G_\alpha \to G$ of geometrically finite groups into G such that

$$\{f_\alpha^*\} : H^i(G;\hat{Z}) \to \prod_\alpha H^i(G_\alpha;\hat{Z})$$

<u>is injective for all</u> $i \in Z$.

<u>4.3</u>. Given a representation $\varrho: G \to GL_m(\mathbb{C})$ of the group G we consider the representations $\varrho_\alpha = \varrho f_\alpha$ of the G_α in Theorem 8. Since Theorem 7 applies to their profinite Chern classes $\hat{c}_j(\varrho_\alpha)$ it immediately follows that one has, for any automorphism σ of $\mathbb{C}$, $\hat{c}_j(\varrho^\sigma) = \hat{k}(\sigma)^j \hat{c}_j(\varrho)$.

<u>Theorem 7'</u>. <u>The statement of Theorem 7 applies to arbitrary groups</u> G .

<u>Remark</u>. In the approximation procedure above a group G is approximated by torsion-free groups G_α . This may appear unnecessarily complicated, e.g., for finite groups. It should, however, be noted that Theorem 7 could by formulated for groups which are <u>virtually</u> of geometrically finite type, thus avoiding any approximation for finite groups.

<u>5</u>. PROOF OF THE MAIN THEOREM

<u>5.1</u>. Let G be an arbitrary group, K a number field, and $\varrho : G \to GL_m(\mathbb{C})$ a representation with all character values in K . Then any $\sigma \in Gal(\mathbb{C}/K)$ leaves the character values fixed, and by Theorem 3 we have $c_j(\varrho^\sigma) = c_j(\varrho)$ for $j = 1,2,3,\ldots$; the same holds for $\hat{c}_j(\varrho) \in H^{2j}(G;\hat{Z})$. From Theorem 7' it follows that

$$\hat{c}_j(\varrho) = \hat{k}(\sigma)^j \hat{c}_j(\varrho)$$

for all $\sigma \in Gal(\mathbb{C}/K)$. This implies, by Proposition 5, that the order of $\hat{c}_j(\varrho)$ is finite and divides $\bar{E}_K(j)$, which proves part A) of the Main Theorem.

5.2. It remains to prove part B). To do this we use the calculations performed in [5]. We recall (Theorem 4.12 of [5]) that $\bar{E}_K(j)$ is the best possible bound for the order of the Chern classes $c_j(\varrho)$ of representations of finite groups defined over the number field K; with the only exception that it is $\frac{1}{2}\bar{E}_K(j)$ if j is even and K formally real. Thus B) will follow if we prove

Theorem 9. *Let K be a formally real number field, and j even > 0. Then there exists a finite 2-group Q and a representation ϱ of Q with character values in K such that*

$$\frac{1}{2}\bar{E}_K(j)\,c_j(\varrho) \neq 0 .$$

Proof. We put $j = 2^{\delta}t$ with t odd, $\delta \geq 1$, and first deal with the case $\delta \geq 2$. Let γ be the integer for which $\zeta_{2^{\gamma+1}} \in K(\sqrt{-1})$, but $\zeta_{2^{\gamma+2}} \notin K(\sqrt{-1})$, i.e., $\gamma = \gamma_K(2)$ in the notation of [5]. Let Q be the generalized quaternion group of order $2^{\gamma+\delta+1}$, and χ a faithful $\mathbb{C}$-irreducible character of Q (of degree 2). The group $\mathrm{Gal}(K(\zeta_{2^{\gamma+\delta}} + \bar{\zeta}_{2^{\gamma+\delta}})/K) = \Gamma$ is cyclic of order $2^{\delta-1}$, cf. [5]. Thus

$$\psi = \sum_{\sigma\in\Gamma} \sigma\chi$$

is the character, with values in K, of a representation of Q of degree 2^{δ}. It is well-known that $c_2(\sigma\chi)$ is a generator of $H^4(Q;\mathbb{Z}) \cong \mathbb{Z}/2^{\gamma+\delta+1}\mathbb{Z}$. From the cohomology ring structure of $H^*(Q;\mathbb{Z})$ it follows that $c_j(t\psi) = c_{2^{\delta}}(\psi)^t$ has maximal possible order $2^{\gamma+\delta+1}$; this is precisely the 2-primary part of $\bar{E}_K(j)$, cf. Prop. 3.4 (d) of [5].

In the case $\delta = 1$, i.e., $j = 2t$ with t odd, we use the above construction of the character ψ of Q for $\delta = 2$. One verifies that $c_j(t\psi)$ has order $2^{\gamma+2}$ which is the 2-primary part of $\bar{E}_K(j)$ (the argument is similar to that in the proof of Prop. 4.11 (b) of [5]).

6. APPENDIX: REMARK ON THE FIRST CHERN CLASS

The universal bound $\bar{E}_K(1)$ is valid for the ordinary Chern class $c_1(\varrho)$ itself, for any group G and any representation $\varrho : G \to GL_n(\mathbb{C})$ with character values in the number field K . To show this we proceed as follows.

We first note that $c_1(\varrho) = c_1(\det \varrho)$ where $\det \varrho$ is the representation of degree 1 given by the determinant of $\varrho(g)$, $g \in G$. If all character values of ϱ are in K then so are the values of $\det \varrho$. Thus $\det \varrho$ factors through the inclusion $\iota : K^* \to GL_1(\mathbb{C}) = \mathbb{C}^*$.

The multiplicative group K^* is the direct product of a free Abelian group and a cyclic group C of order $\bar{E}_K(1)$. It easily follows that $c_1(\iota)$ is of order $\bar{E}_K(1)$. This implies that $c_1(\det \varrho) = c_1(\varrho)$ is of finite order dividing $\bar{E}_K(1)$.

REFERENCES

[1] G. Baumslag, E. Dyer and A. Heller: The topology of discrete groups. J. Pure Appl. Algebra 16 (1980), 1-47.

[2] N. Bourbaki, Algèbre; Hermann, Paris, 1958.

[3] P. Cassou-Noguès: Valeurs aux entiers négatifs des fonctions zêta et fonctions zêta p-adiques. Inventiones math. 51 (1979), 29-59.

[4] P. Deligne and D. Sullivan: Fibrés vectoriels complexes à groupe structural discret. C.R. Acad. Sc. Paris, t. 281, Série A, (1975), 1081-1083.

[5] B. Eckmann and G. Mislin: Chern classes of group representations over a number field. Compositio Mathematica 44 (1981), 41-65.

[6] A. Grothendieck: Classes de Chern et représentations linéaires des groupes discrets. Dans: Dix exposés sur la cohomologie des schémas, Amsterdam, North-Holland 1968.

[7] G. Mislin: Classes caractéristiques pour les représentations des groupes discrets; Séminaire Dubreil-Malliavin Paris 1981, Lecture Notes in Math., Springer-Verlag, Vol. 924.

[8] J.-P. Serre: Cohomologie des groupes discrets. Annals of Math. Studies 70 (1971), 77-169.

[9] C. Soulé: Classes de torsion dans la cohomologie des groupes arithmétiques. C.R. Acad. Sc. Paris, t. 284, Série A (1977), 1009-1011.

[10] D. Sullivan: Genetics of homotopy theory and the Adams conjecture. Annals of Math. 100 (1974), 1-79.

AUTOMORPHISMS OF SURFACES AND CLASS NUMBERS: AN ILLUSTRATION OF THE G-INDEX THEOREM

John Ewing
Indiana University, Bloomington, Indiana 47405, USA

Introduction

This brief essay is an illustration of one simple application of the Atiyah-Bott-Segal-Singer G-Index Theorem. It is not intended for experts — if you know all about the Index Theorem, stop reading now — but it *is* intended for Algebraic topologists. The purpose is to show in a non-technical setting how one applies the G-Index Theorem, why it is so effective, and how it can lead to surprising connections with Number Theory.

The specific problem we ask is quite elementary, involving automorphisms of Riemann surfaces. It is the pattern of application which is important: the G-Index Theorem allows us to compute algebraic invariants which are otherwise exceedingly difficult to calculate; the algebraic questions one then asks about the invariants are often already answered by number theorists. This is a pattern which repeats itself in much more complicated settings.

We begin with a digression on a subject in Number Theory which has frequently arisen in Topology in recent years — the ideal class number.

The Class Number

The ideal class number is an invention of Kummer; its purpose was to fix-up a faulty proof of the Fermat conjecture: $x^p + y^p = z^p$ has no non-trivial solutions for any odd prime p.

The faulty proof goes like this. We first factor $x^p + y^p$. Of course, to do this we must work over the ring $\mathbb{Z}[\lambda]$ where λ is a p-th root of unity. We have

$$(x+y)(x+\lambda y)\cdots(x+\lambda^{p-1}y) = z^p .$$

Next we can show (with a little effort) that the $x + \lambda^i y$ are relatively prime. Factoring both sides into primes, we conclude that $x + \lambda^i y$ is a p-th power in $\mathbb{Z}[\lambda]$. Finally, one derives a contradiction; it's not important what contradiction — it's similar to the fact that in the ordinary integers there are no consecutive p-th powers.

Of course, this is wrong: the ring $\mathbb{Z}[\lambda]$ does not have unique factorization in general and so one cannot conclude that the $x + \lambda^i y$ are p-th powers. Kummer's idea was to correct the flaw by adding ideal numbers — we'll use the more modern terminology of ideals due to Dedekind.

An ideal is a subset of $\mathbb{Z}[\lambda]$ which is closed under addition and closed under multiplication by elements of $\mathbb{Z}[\lambda]$. Obvious example: take any $a \in \mathbb{Z}[\lambda]$ and consider all multiples of a. That's a principal ideal denoted $\langle a \rangle$. The problem is that not all ideals are principal.

Now Kummer showed that ideals do factor uniquely. The previous argument can be repeated, using ideals rather than numbers. We factor:

$$\langle x+y\rangle\langle x+\lambda y\rangle\cdots\langle x+\lambda^{p-1}y\rangle = \langle z\rangle^p .$$

The ideals are still relatively prime. We can factor both sides into the product of primes (ideals!), and we can correctly conclude that each $\langle x+\lambda^i y\rangle$ is a p-th power. Alas, each is a p-th power of an ideal, not a number. Our previous contradiction has vanished — or has it?

To arrive at the same contradiction as before we need a condition which will ensure that whenever a principal ideal is the p-th power of some ideal it must be the p-th power of a principal ideal. This is Kummer's first great idea. Look at the collection of all ideals modulo the principal ideals — that's a group called the ideal class group. It has finite order (a theorem) and its order h is called the ideal class number. Our problem is solved if no element in the ideal class group has order p; that is, if $p \nmid h$. Then each $\langle x+\lambda^i y\rangle$ is the p-th power of an ideal, necessarily principal, and we arrive at the same contradiction as before. We say p is regular if $p \nmid h$; the Fermat conjecture is thus proved for all regular primes.

Clearly this would be an exercise in semantics if there were no way to determine when a prime was regular. This was Kummer's second great idea: he provided a simple and effective way to do this.

The class number is extremely difficult to compute and is presently only known for $p < 67$. But it always factors into two integers, $h = h^- h^+$, and h^-, called the first factor, is easy (in principle) to compute. Indeed,

h^- <u>is the determinant of a certain</u> $(p+1)/2 \times (p+1)/2$ <u>integer matrix</u>

Warning: While h^- can be computed in principle, it grows very rapidly. For $p < 23$, $h^- = 1$; for $p = 23$, $h^- = 3$; but for $p = 199$, $h^- \approx 10^{33}$.

For his special purposes Kummer devised a special criteria for regularity, thus circumventing the difficulty of computing h. (He showed p is regular iff p divides the numerator of one of the first $(p-1)/2$ Bernoulli numbers.) Nonetheless, the computation of h and h^- has been a central theme in algebraic number theory for over a century.

<u>Automorphisms of Surfaces</u>

Now we can return to Topology...well, almost. We can work our way back to the present by starting with Riemann, one of the fathers of Topology.

It is remarkable that in the same year (1851) that Kummer began publishing his important work, Riemann delivered his famous address on Riemann surfaces. Of course, a Riemann surface is an analytic object, but (surprisingly) many of its properties depend only on its topology. Towards the end of that address Riemann makes a profound observation: While Riemann surfaces are constructed in order to study multivalued maps on the complex plane, the interesting objects of study are the automorphisms (holomorphic homeomorphisms) of a given surface.

It was a fruitful remark, one that has led to many beautiful and surprising results. For example, if a surface S has genus > 1 then the automorphism group of S is finite! That means that for most surfaces automorphisms have finite period.

What kinds of automorphisms are there of a fixed period p? (To make life simple, we assume p is prime.) That's a vague question which we need to make more precise. We can make it more precise by assigning an algebraic invariant to each automorphisms — that's certainly in the spirit of Algebraic Topology.

Now one studies a Riemann surface S by studying the holomorphic and meromorphic functions and differentials on S. We can define our invariant by considering the holomorphic differentials. Of course, we can add two differentials and multiply by complex scalars; that is, the set of holomorphic differentials on S forms a vector space V. The dimension of V is precisely the genus of S; it doesn't depend on the analytic structure at all.

An automorphism T of period p induces a linear mapping T^* on V. (If we think of a differential ω as a section of the cotangent bundle then $T^*(\omega) = \omega \circ T^{-1}$.) Of course, T^* also has order p and so defines a representation of the cyclic group of order p. To study T^* we can study the character of this representation; that is, we consider the trace $\mathrm{Tr}(T^*)$. If T^* is a linear mapping of period p then the eigenvalues of T^* are powers of a p-th root of unity λ. Hence, $\mathrm{Tr}(T^*) \in \mathbb{Z}[\lambda]$. We can at last ask our realization question more precisely.

<u>Question</u>: For a fixed prime p, which elements of $\mathbb{Z}[\lambda]$ can be realized as $\mathrm{Tr}(T^*)$ for some automorphism T of period p on some Riemann surface S? That is, which characters can be realized?

There is one easy restriction, known even to Riemann. It is a basic result of complex analysis (the "Hodge decomposition") that $V \oplus \bar{V} = H^1(S;\mathbb{C}) = H^1(S;\mathbb{Z}) \otimes \mathbb{C}$. It follows that $\mathrm{Tr}(T^*) + \overline{\mathrm{Tr}(T^*)}$ must be the trace of an integer matrix and hence must be an integer itself. It's easy to see that many elements of $\mathbb{Z}[\lambda]$ fail to satisfy this. Indeed, if we set

$$A = \{x \in \mathbb{Z}[\lambda] \mid x + \bar{x} \in \mathbb{Z}\}$$

then we see that A is a free $\mathbb{Z}$-module generated by 1, $\lambda - \lambda^{-1}$, $\lambda^2 - \lambda^{-2}, \ldots, \lambda^m - \lambda^{-m}$, where $m = (p-1)/2$.

Suppose we let B denote the group of elements of A which are realizable. The question is: Does $B = A$? If not, what is the index.

The G-Index Theorem

The realizability question of the previous section seems, at first glance, to be hopeless. It is extremely difficult to directly compute Tr(T*) for a given automorphism T. The difficulty is not in producing such automorphisms — we can produce them in great variety — but rather in determining the behavior of T* on the space of holomorphic differentials. How then can we hope to compute all possible traces for all possible automorphisms of a fixed period? The answer is: use the G-Index Theorem or, more precisely, a special case of it known as the Eichler Trace Formula [3; p. 264].

The Index Theorem (without the G) always says the same thing: the analytic index = the topological index. It allows us to compute certain analytic invariants in terms of topological information. The great strength of the theorem is that the indices take on many different meanings in many different contexts. For example, in our special context the Index Theorem says that the dimension of the space of holomorphic differentials (an analytic invariant) is precisely the genus of the surface (a topological invariant).

The G-Index Theorem provides us with a similar tool when the situation we are studying comes equipped with an action of a group G. It allows us to compute the value of an analytically defined character of the group G in terms of topological information.

In our special case the value of the character is simply Tr(T*). How do we determine it in terms of topological information?

Suppose T is an automorphism of period p. Then T will fix a (possibly empty!) set of isolated points $P_1, P_2, \dots, P_t$. (Remember that T must be orientation preserving if it is holomorphic.) We can describe the "type" of each fixed point by examining the differential of T acting on the tangent space at that point. If λ is a fixed p-th root of unity then dT acting on the tangent space at P_i must be multiplication by some power of λ, say λ^{k_i}. We say that P_i has type λ^{k_i}. The collection of fixed points together with their types is called the "fixed point data" of T.

Now the G-Index Theorem (in this special case) tells us how to determine Tr(T*) in terms of the fixed point data. Indeed the formula is simply:

$$\mathrm{Tr}(T^*) = 1 - \sum_{P_i} \frac{1}{\lambda^{k_i} - 1}$$

(This is in fact a special case of the Atiyah-Bott fixed point formula [1; p. 459, ex. 1] which is in turn a special case of the G-Index Theorem. It is, however, actually a result due to Eichler around 1930 and independently proved by Chevalley and Weil in 1934 [2]. A good reference for an "elementary" proof is [3; p. 256 ff.].)

The G-Index Theorem allows us to do what seemed to be impossible: We can compute $\mathrm{Tr}(T^*)$ for an automorphism in terms of easily determined, topological data.

Realizing Fixed Point Data

Our problem is nearly solved. We must merely determine which sets of fixed point data can be realized, and then solve the algebraic problem of determining which numbers of A can be realized by the formula for $\mathrm{Tr}(T^*)$.

Realizing fixed point data is a simple problem in covering spaces. Suppose we try to produce an automorphism T of period p on a surface S with fixed points $P_1,\ldots,P_t$ of types $\lambda^{k_1},\ldots,\lambda^{k_t}$.

Our first observation is that it is enough to produce a diffeomorphism; it will be an automorphism for some complex structure.

Now we also observe that after deleting the fixed points, S is a p-fold covering of its orbit space and T is just a deck transformation. We can therefore try to build S from its orbit space. Start with t deleted disks with boundary curves $C_1, C_2,\ldots,C_t$. (These will be the deleted neighborhoods of our fixed points in the orbit space.) As usual, join the C_i by paths to a common base point. To create S we must first take a p-fold covering of this configuration — such a covering is classified by a homomorphism ϕ from the fundamental group into $\mathbb{Z}/p\mathbb{Z}$. If we identify the deck transformation T as a generator for $\mathbb{Z}/p\mathbb{Z}$, it is not hard to see that P_i has type λ^{k_i} precisely when $\phi(C_i^{k_i}) = T$.

The key observation is that we can complete our construction of S and T if and only if the curve $C_1 C_2 \ldots C_t$ lifts to a disjoint union of p circles. We can then cap-off each circle and extend T —

it simply permutes the caps. This is true iff $\phi(C_1 \ldots C_t) = 0$ or, what is the same thing, $\sum 1/k_i \equiv 0 \bmod p$. This is a necessary and sufficient condition for realizing a set of fixed point data.

The remaining algebraic problem is straight-forward but messy. Recall that A (the set of all numbers in $\mathbb{Z}[\lambda]$ satisfying the obvious necessary condition to be a trace) is a free $\mathbb{Z}$-module generated by $1, \lambda - \lambda^{-1}, \ldots, \lambda^m - \lambda^{-m}$, where $m = (p-1)/2$. We need to determine the index of the submodule B of numbers of the form

$$1 - \sum \frac{1}{\lambda^{k_i} - 1}$$

where

$$\sum 1/k_i \equiv 0 \bmod p .$$

This is a familiar kind of problem to algebraists and simply requires finding (in principle) a basis for B ; the index of B in A is then a determinant. In this case it is, in fact, a familiar determinant. Indeed, it is the <u>same</u> determinant we mentioned in connection with the class number:

the index of B in A is h^-,

the first factor of the class number!

Now the wonderful consequence is that all the work that number theorists have done for a century can be applied to our question about automorphisms. For small primes $(< 23) h^- = 1$ and so A and B are the same; there are no restrictions on which automorphisms can be realized other than the obvious ones. For larger primes, however, A and B are quite far apart (recall that $h^- \approx 10^{33}$ for $p = 199$); there seem to be many more restrictions on automorphisms of period p — restrictions which are a little mysterious.

Why is there any connection between the class number and automorphisms of Riemann surfaces? Is it purely surperficial — an accident of computation — or is there some deeper, underlying reason? Unknown.

For our purposes it is enough to point out that there is a connection and that it is made possible by the G-Index Theorem. The "significance" is that to answer our question we need not do any more work — the nubmer theorists have done it for us!

References

[1] Atiyah, M.F. and Bott, R. (1968). A Lefschetz fixed point formula for elliptic complexes: II. Applications. Ann. of Math., 88, 451-491.

[2] Chevalley, C. and Weil, A. (1934). Uber das Verhalten der Integrale 1. Gattung bei Automorphismen des Funktionenkorpers. Abh. aus dem Math. Sem. der Univ. Hamburg, 10, 358-361.

[3] Farkas, H.M. and Kra, I. (1980). Riemann Surfaces. New York: Springer-Verlag.

THE REAL DIMENSION OF A VECTOR BUNDLE AT THE PRIME TWO

Edgar H. Brown, Jr. and Franklin P. Peterson[1]

ABSTRACT

Suppose ξ is a real, stable vector bundle over a CW complex X of dimension less than 2n+1. It is shown that the obstructions to realizing ξ as an n-plane bundle may be arranged so as to be a twisted Bockstein on w_n, when n is even, and cosets of $H^*(X;Z_2)$.

INTRODUCTION

In this paper we show that if ξ is a real, stable vector bundle, then the obstructions to realizing ξ by an n-plane bundle are 2-primary in dimensions less than 2n. Under the assumption that ξ is oriented, the results of this paper were obtained by Copeland and Mahowald in [5]. See also the related results of Glover and Mislin [7].

Let

$$BO_n \xrightarrow{f_k} BO_n^{(k)} \xrightarrow{p_k} BO$$

be the kth stage in the Postnikov system for the projection p: $BO_n \to BO$, that is:

$$(f_k)_*: \ \pi_i(BO_n) \approx \pi_i(BO_n^{(k)}), \ i \leq k$$

$$(p_k)_*: \ \pi_i(BO_n^{(k)}) \approx \pi_i(BO), \ i > k.$$

Theorem 1.1.

There is a tower of fibrations

$$B_N \to \dots \to B_j \to B_{j-1} \to \dots B_0 = BO$$

and a map $F: B_N \to BO_n^{(2n-2)}$ such that

(i) If $\bar{p}: B_N \to B_0$ is the composition of the above projections, then $p_{2n-2}F = \bar{p}$. (F is a 2 local homotopy equivalence.)

(ii) Except for n even and $j = 1$, $B_j \to B_{j-1}$ has fibre $K(Z_2, n_j)$.

(iii) If n is even, $B_1 \to B_0$ has fibre $K(Z_{(2)}, n)$ and k-invariant $\delta_t^* w_n \in H^*(BO; Z_{(2)}^t)$, where w_n is the nth Stiefel-Whitney class, $Z_{(2)}$ is the rational numbers with odd denominators, $Z_{(2)}^t$ denotes local coefficients with respect to w_1 and δ_t^* is the twisted Bockstein operation associated to $Z_{(2)} \xrightarrow{2} Z_{(2)} \to Z_2$ (see below for definition of δ_t^*).

In [2] we constructed a space BO/I_n over BO and conjectured that the projection $BO/I_n \to BO$ lifts to $BO_{n-\alpha(n)}$; this conjecture implies that every smooth n-manifold immerses in $R^{2n-\alpha(n)}$. Theorem 1.1 and the following theorem show that one need only consider obstructions in $H^*(BO/I_n; Z_2)$. This fact is used by R. Cohen ([4]) in his recent proof of the immersion conjecture.

Theorem 1.2.

If $n-\alpha(n)$ is even, then $\delta_t^* w_{n-\alpha(n)} \in H^*(BO/I_n; Z_{(2)}^t)$ is zero.

One may recast 1.1 in a more obstruction theoretic form as follows:

Corollary 1.3.

Suppose ξ is a k-plane bundle over a finite CW complex X with $\dim X < k$, $\dim X < 2n$, $H^q(X; Z_2) = 0$ for $q > n+1$, and if n is odd, $w_{n+1}(\xi) = 0$, or if n is even, $\delta_t^* w_n(\xi) = 0$. Then $\xi = \xi' + 0^{k-n}$, where ξ' is an n-plane bundle and 0^{k-n} is the trivial (k-n)-plane bundle.

LOCALIZATION OF A MAP

In this section we recall the definitions of a local system of groups, cohomology with local coefficients, the Serre spectral sequence with local coefficients, and localization of a map.

Suppose X is a space. Let P(X) be the category whose objects are points of X and morphisms between x_0 and x_1 are the homotopy class of paths from x_0 to x_1. A local system of groups G, on X, is a covariant functor from $P(X)$ to the category of groups. The cohomology groups, $H^q(X;G)$ are constructed using cochains which assign to each q-simplex, $T: \Delta_q \to X$, an element of $G(T(d_0))$ where d_0 is the 0th vertex of Δ_q (see [6]).

Suppose $F \to E \xrightarrow{p} B$ is a Hurewicz fibration with lifting function λ, that is, for each pair (e,α), $e \in E$, $\alpha: I \to B$, such that $\alpha(0) = pe$, λ gives an element $\lambda(e,\alpha)$ in $p^{-1}(\alpha(1))$ and λ is continuous in e and α. Suppose G is a local system of abelian groups on B. We obtain a local system $H^q(F;G)$ as follows: for $b \in B$,

$$H^q(F;G)(b) = H^q(p^{-1}(b), G(b));$$

and if $\alpha: I \to B$

$$H^q(F;G)(\alpha) = G(\alpha)_*(\lambda_\alpha^*)^{-1}$$

where $\lambda_\alpha: p^{-1}(\alpha(0)) \to p^{-1}(\alpha(1))$ is defined by $\lambda_\alpha(e) = \lambda(e,\alpha)$.

Theorem 2.1.

If G is a local system of groups on B and $F \to E \xrightarrow{p} B$ is a Hurewicz fibration, then the Serre spectral sequence for $H^*(E;p^*G)$ has its E_2 term given by

$$E_2^{p,q} = H^p(B;H^q(F;G)).$$

Suppose $p: E \to B$ is a fibration with simply connected fibre and suppose

$$E \to \ldots \to E^{(k)} \to E^{(k-1)} \to \ldots \to E^{(1)} = B$$

is the Postkinov system of p. The space $E^{(k)}$ is constructed from $E^{(k-1)}$ as follows: let $\pi = \pi_1(B)$, $\pi_k = \pi_k(B)$, W the universal cover of $K(\pi,1)$ and P_k the space of paths in $K(\pi_k,k+1)$ starting at the base point. The action of π on π_k gives an action of π on $K(\pi_k,k+1)$ and P_k. Then $E^{(k)}$ is the induced fibration in

$$\begin{array}{ccc} E^{(k)} & \longrightarrow & W \times_\pi P_k \\ {\scriptstyle K(\pi_k,k)}\downarrow & & \downarrow \\ E^{(k-1)} & \xrightarrow{\theta^{k+1}} & W \times_\pi K(\pi_k,k+1) \end{array}$$

where θ^{k+1} induces an isomorphism on π_1 and may be viewed as the k-invariant lying in $H^{k+1}(E^{(k-1)};\pi_k^t)$, the t indicating local coefficients.

Suppose R is a subring of the rationals, with unit. We construct, by induction, a tower of spaces and maps:

$$\begin{array}{ccccccc} E^{(k)} & \longrightarrow & E^{(k-1)} & \to \cdots \longrightarrow & E^{(1)} = B \\ \downarrow g_k & & \downarrow & & \downarrow \\ E^{(k)} \otimes R & \to & E^{(k-1)} \otimes R & \to \cdots \to & E^{(1)} \otimes R = B \end{array}$$

where the kth fibre is $K(\pi_k \otimes R,k)$ and g_k induces an isomorphism,

$$H^q\big((E^{(k)} \otimes R);G \otimes R\big) \approx H^q\big(E^{(k)};G \otimes R\big)$$

for any local system G on $E^{(k)}$ induced from a local system on B. Suppose E^{k-1} and g_{k-1} have been constructed. Let $\theta^{k+1} \otimes R \in H^{k+1}\big(E^{(k-1)} \otimes R;\pi_k \otimes R\big)$ correspond under g_{k-1} to $j_*\theta^{k+1} \in H^{k+1}\big(E^{(k-1)};\pi_k \otimes R\big)$, where $j\colon \pi_k \subset \pi_k \otimes R$. Then $E^{(k)} \otimes R$ is the fibration over $E^{(k-1)} \otimes R$ induced by $\theta^{k+1} \otimes R$. The existence of g_k follows by construction and the fact that it induces the appropriate cohomology isomorphism follows from 2.1 and

$$H^*\big(K(\pi \otimes R,h),G \otimes R)\big) \approx H^*\big(K(\pi,h),G \otimes R\big)$$

for any group G.

PROOF OF 1.1.

Let F_n be the fibre of $BO_n \to BO$. Note F_n is also the fibre of $BSO_n \to BSO$. We first consider the case n odd. Then

$$H^q(BSO;Z[1/2]) \approx H^q(BSO_n;Z[1/2])$$

for $q < 2n+2$, since the Pontrjagin class $P_{n+1/2}$ is the first class in BSO which vanishes on BSO_n. Hence for $i \leq 2n$, $\pi_i(F_n)$ is a finite group whose order is a power of two. The groups $\pi_i(F_n)$, $i \leq 2n$, are modules over $Z[\pi_i(BO)] = Z[Z_2]$ and such modules, groups of order a power of two, are nilpotent. Hence by results in [1], $BO_n^{(2n)} \to BO$ may be factored into a sequence of fibrations with fibre $K(Z_2,n_i)$. We take $B_N = BO_n^{(2n-2)}$ and F = identity. This proves 1.1 for n odd.

Suppose n is even. Note $F_n \to F_{n+1}$ and $BO_n \to BO_{n+1}$ have the same fibre, namely, S^n. Hence $\pi_i(F_n) = Z$ for $i = n$ and is a finite group for $n < i \leq 2n$. In [3] it is shown that the first obstruction to a section of $BO_n \to BO$, when n is even, is $\delta_t^* w_n \in H^{n+1}(BO; Z^t)$, where δ_t^* is the twisted Bockstein defined as follows: suppose $V \in C^n(BO; Z)$ represents $v \in H^n(BO; Z_2)$, $\delta V = 2U$, $W_1 \in C^1(BO; Z)$ represents w_1 and W_1 only takes the values 0 or 1 on each simplex. Then, one defines δ_t by

$$\delta_t V = \delta V - 2W_1 \cup V.$$

(One may easily check that this is the usual local coefficient boundary operator.) Note

$$\delta_t V = 2(U - W_1 \cup V).$$

We define

$$\delta_t^* v = \{U - W_1 \cup V\} \in H^{n+1}(BO; Z^t).$$

Consider the following commutative diagram:

$$\begin{array}{ccccccc}
BO_{n-1} \to BO_n^{(2n-2)} & \longrightarrow & B_1 & \longrightarrow & BO_n^{(2n-2)} \otimes Z[1/2] \\
\downarrow & & \downarrow & & \downarrow p_1' \\
B_2 & \longrightarrow & BO_n^{(n)} & \to & BO_n^{(n)} \otimes Z[1/2] \\
\downarrow & & \downarrow & & \downarrow p_1 \\
BO_n^{(2n-2)} \otimes Z_{(2)} & \xrightarrow{p_2'} & BO_n^{(n)} \otimes Z_{(2)} & \xrightarrow{p_2} & BO
\end{array}$$

where $BO_n^{(k)}$ and $BO_n^{(k)} \otimes R$ come from $BO_n \to BO$ as in section two, and all the squares not containing BO are pullbacks. The homotopy groups of the fibres of p_1' and p_2' have order odd and a power of two, respectively. If ρ_1: $Z \to Z[1/2]$ and ρ_2: $Z \to Z_{(2)}$ are the inclusions and δ_t^* is the twisted Bockstein associated to w_1 and $Z \to Z \to Z_2$, then the k-invariants of p_1 and p_2 are $\rho_1 \delta_t^* w_n$ and $\rho_2 \delta_t^* w_n$, respectively.

To prove 1.1, n even, we take $B_N = BO_n^{(2n-2)} \otimes Z_{(2)}$. Then $B_N \to BO_n^{(n)} \otimes Z_{(2)}$ factors into fibrations with fibres $K(Z_2, n_i)$ as in the n odd case above. Hence $B_N \to BO$ has the desired properties and it remains to show that $BO_n^{(2n-2)} \to B_N$ has a section. Let $B = BO_n^{(n)} \otimes Z_{(2)}$. It is sufficient to show $B_1 \to B$ has a section.

We first show that p_1p_1' has a section. Consider the commutative diagram:

$$\begin{array}{ccc} BO_{n-1}^{(2n-2)} \otimes Z[1/2] & \rightarrow & BO_n^{(2n-2)} \otimes Z[1/2] \\ & q \searrow \quad \swarrow p_1p_1' & \\ & BO & \end{array}$$

Since n-1 is odd, $\pi_i(F_{n-1})$ is a 2-group for $i \leq 2n-2$. Hence q is a homotopy equivalence and thus p_1p_1' has a section.

By construction $\rho_2\delta_t^* w_n = 0$ in $H^*(B)$ and, since p_1 has a section, $\rho_1\delta_t^* w_n = 0 \in H^*(B)$. Suppose $V \in C^{n+1}(B;Z)$ is a cochain representing $\delta_t^* w_n$: then

$$V = \frac{1}{q}(\delta_t V_1) = \frac{1}{r}(\delta_t V_2)$$

where $V_i \in C^{n-1}(B;Z)$, q is odd and r is a power of two. There are integers a and b so that $aq + br = 1$, and hence

$$V = \delta_t(aV_1 + bV_2).$$

Therefore $\delta_t^* w_n = 0$ in B and $BO_n^{(n)} \rightarrow B$ has a section. This section together with the section of p_1, from the previous paragraph, give two liftings of $B \rightarrow BO$ to $B \rightarrow BO_n^{(n)} \otimes Z[1/2]$ which will differ by an element in $H^n(B;Z[1/2]^t)$. The Serre spectral sequence for $K(Z_{(2)},n) \rightarrow B \rightarrow BO$ and $H^*(B;Z[1/2]^t)$ has, by 2.1, its E_2 given by

$$E_2^{p,q} = H^p\left(BO;H^q\left(K(Z_{(2)},n);Z[1/2]^t\right)\right).$$

But $H^q\left(K(Z_{(2)},n);Z[1/2]\right) = 0$ for $q > 0$. Hence $E_2^{p,q} = 0$ for $q > 0$ and $E_2^{p,0} = H^p\left(BO;Z[1/2]^t\right)$. By the Thom isomorphism theorem,

$$H^p(BO;Z[1/2]^t) \approx H^p(MO;Z[1/2]) = 0.$$

Hence $H^n(B;Z[1/2]^t) = 0$, the two liftings above are the same and, since $B \rightarrow BO_n^{(n)} \otimes Z[1/2]$ lifts to $BO_n^{(2n-2)} \otimes Z[1/2]$, $B_1 \rightarrow B$ has a section and the proof of 1.1 is complete.

PROOF OF 1.2.

Let MO/I_n denote the Thom spectrum of $BO/I_n \to BO$ and let $U \in H^0(MO/I_n;Z_2)$ and $U^t \in H^0(MO/I_n;Z^t)$. The Thom isomorphism for local coefficients has the form:

$$\cup_t U^t\colon\ H^q(BO/I_n;Z^t) \approx H^q(MO/I_n;Z)$$

where $\cup_t$ is defined as follows: if $u,v \in C^*(BO/I_n;Z)$,

$$u \cup_t v(A_0,\dots,A_{p+q}) = (-1)^{w_1(A_0A_p)}u(A_0,\dots,A_p)v(A_p,\dots,A_{p+q}).$$

We next show that

$$(\delta_t^* w_k) \cup_t U^t = \delta^*(w_k \cup U).$$

One may check that

$$\delta(u \cup_t v) = \delta_t u \cup_t v + (-1)^{|u|}u \cup_t \delta_t v.$$

Let $\rho\colon\ Z \to Z_2$ be the quotient map. Let $u,v \in C^*(BO/I_n;Z)$ represent w_k and U^t, respectively. Then $\rho(u \cup v) = \rho(u \cup_t v)$ and $\delta_t v = 0$. Hence

$$\begin{aligned}\delta^*(w_k \cup U) &= \{\delta(u \cup_t v)/2\}\\ &= \{(\delta_t u/2) \cup_t v\}\\ &= (\delta_t^* w_k) \cup_t U^t.\end{aligned}$$

Let $k = n - \alpha(n)$ and suppose k is even. To prove 1.2, namely, that $\delta_t^* w_k = 0$ in $H^*(BO/I_n;Z^t)$, we may show that $\delta^*(w_k U) = 0$ in $H^*(MO/I_n;Z)$. Since $w_{k+1} \in I_n$ (see [2]), $Sq^1(w_k U) = w_{k+1}U = 0$. We show that $w_k U = Sq^1(y)$ and hence $\delta^*(w_k U) = 0$ because $\delta^* Sq^1 = 0$.

Let $\{u_\omega\}$ be a basis for $H^*(MO;Z_2)$ over A = Steenrod algebra. The set $\{\chi(Sq^I)u_\omega \mid I \text{ admissible}\}$, χ the canonical antiautomorphism of A, is a Z_2 basis for $H^*(MO;Z_2) = H^*(MO)$.

$$H^*(MO/I_n) = H^*(MO)/J$$

where J is the Z_2 subspace generated by $\chi(Sq^I)u_\omega$ where the first entry of

I, i_1, satisfies $2i_1 > n - |u_\omega|$. Hence

$$w_k U = Sq^k U = \Sigma \quad \chi(Sq^{I_j})U$$

and $Sq^1\chi(Sq^{I_j})U \in J$. If I_j has 1 as its last entry, $Sq^1\chi(Sq^{I_j}) = 0$ and $\chi(Sq^{I_j}) = Sq^1\chi(Sq^I)$. Otherwise, $Sq^1\chi(Sq^{I_j}) = \chi\left(Sq^{(I_j,1)}\right)$ and $\chi(Sq^{I_j})U \in J$. Hence $w_k U = Sq^1 y$ and the proof of 1.2 is complete.

Brandeis University

Massachusetts Institute of Technology

REFERENCES

1. A.K. Bousfield and D.M. Kan; Homotopy limits, completions and localizations, Springer-Verlag, Lecture Notes in Mathematics 304 (1972).

2. E.H. Brown and F.P. Peterson; "A universal space for normal bundles of n-manifolds", Comment. Math. Helv. 54 (1979), 405-430.

3. N. Steenrod; The Topology of Fibre Bundles, Princeton University Press, 1951.

4. R. Cohen; "The immersion conjecture for differentiable manifolds", to appear.

5. A.H. Copeland, Jr., and M. Mahowald; "The odd primary obstructions to finding a section in a V_n bundle are zero", Proc. A.M.S. 19 (1968), 1270-1272 .

6. E. Spanier; Algebraic Topology, McGraw Hill, 1966.

7. H.H. Glover and G. Mislin; "Immersions in the metastable range and 2-localization", Proc. A.M.S. 29 (1971), 190-196.

NOTES

1. The work in this paper was supported by NSF grants.

MAPS BETWEEN CLASSIFYING SPACES, III

J.F. Adams and Z. Mahmud

§1. Introduction. Let G and G' be compact connected Lie groups, and let $f: BG \longrightarrow BG'$ be a map. It is shown in [2] that the induced homomorphism $f^*: K(BG) \longleftarrow K(BG')$ carries the representation ring $R(G') \subset K(BG')$ into the representation ring $R(G) \subset K(BG)$. Moreover the induced map $R(G) \longleftarrow R(G')$ can also be induced by a homomorphism $\theta: T \longrightarrow T'$, where T, T' are the maximal tori in G, G'. Here of course one has to state that the behaviour of θ with respect to the Weyl groups W, W' is such that $\theta^*: R(T) \longleftarrow R(T')$ does indeed carry $R(G') \subset R(T')$ into $R(G) \subset R(T)$; in [2] maps θ with this behaviour are called "admissible maps".

Our present purpose is to see what further information can be obtained by using real and symplectic K-theory. Let $RO(G) \subset R(G)$ be the subgroup generated by real representations; similarly for $RSp(G) \subset R(G)$, using symplectic representations. Assume that the group G is semi-simple.

Proposition 1.1. For any map $f: BG \longrightarrow BG'$, the induced homomorphism $R(G) \longleftarrow R(G')$ preserves real elements, in the sense that it carries $RO(G')$ into $RO(G)$; similarly it preserves symplectic elements, in the sense that it carries $RSp(G')$ into $RSp(G)$.

The main problem, however, is to take this result and deduce useful, explicit conclusions about $\theta: T \longrightarrow T'$. For this purpose we must recall some representation-theory. In particular, if ρ is any irreducible (complex) representation of G, then under ρ any element z in the centre Z of G acts as a scalar. If $z^2 = 1$ in G, then of course the scalar $\rho(z)$ is ± 1.

Lemma 1.2 (after Dynkin [5]). (a) Let G be a compact connected Lie group. Then there is a canonical element $\delta \in Z \subset G$ with the following properties. (i) $\delta^2 = 1$. (ii) For any irreducible self-conjugate representation ρ of G, $\rho(\delta)$ is +1 or -1 according as ρ is real or symplectic. (b) More explicitly, when G is simple δ is as follows. If G is SU(n) with n odd, Spin(m) with $m \equiv 0,1,2$ or 7 mod 8, G_2, F_4, E_6 or E_8 then $\delta = 1$. If G is SU(n) with n even or Sp(n) then δ is the matrix -1. If G is Spin(m) with $m \equiv 3,4,5$ or 6 mod 8 then δ is the non-trivial element in the kernel of Spin(m) $\longrightarrow$ SO(m). If G is E_7 then δ is the unique non-trivial element in the centre Z. (c) If $\zeta \in Z$ is any element of the centre such that $\zeta^4 = 1$, then $\delta\zeta^2$ also has properties (i) and (ii) of part (a).

We shall write $I \subset Z$ for the subgroup of elements ζ^2, where $\zeta \in Z$ and $\zeta^4 = 1$; and we shall regard I as an "indeterminacy" which affects elements satisfying (i) and (ii) of part (a).

Our object is now to conclude that if we take f: BG $\longrightarrow$ BG' and pass to an associated admissible map θ: T $\longrightarrow$ T', then $\theta\delta \equiv \delta'$ mod I'; in other words, θ preserves the Dynkin element (modulo indeterminacy). It is apparent from Lemma 1.2 that this condition is sufficient for θ^*: R(G) $\longleftarrow$ R(G') to preserve real and symplectic elements; we aim to prove that this condition is also necessary, at least in certain cases.

This calls for two comments. First, we wish to emphasise that the condition "$\theta\delta \equiv \delta'$ mod I'" is indeed useful and explicit. For example, consider the case G = G' = Sp(1); the centre Z = Z' is $\{\pm 1\}$, and the Dynkin element $\delta = \delta'$ is -1. The appropriate maps θ are those of the form $\theta_k(t) = t^k$, so the condition "$\theta\delta \equiv \delta'$ mod I'" becomes "k is odd".

Secondly we wish to emphasise that further assumptions will be needed. For example, in the case $G = G' = Sp(1)$ we know [6,7] that the correct conclusion is "k is odd <u>or zero</u>", and we know that the case $k = 0$ can actually occur (take the map $f: BG \longrightarrow BG'$ to be constant at the base-point). So in this case we shall need some assumption to exclude the case $k = 0$.

In fact, in general $\theta\delta$ need not even lie in the centre Z' of G'. (For an example, consider the usual injection $Sp(n) \longrightarrow Sp(n+1)$.) However, we have the following result.

<u>Proposition 1.3</u>. Let $f: BG \longrightarrow BG'$ correspond to an admissible map $\theta: T \longrightarrow T'$ which is irreducible in the sense of [2] p20; then θ carries Z into Z'.

Somewhat weaker assumptions should suffice to prove that $\theta\delta$ lies in Z'; for it is sufficient to verify that $\theta\delta$ lies in the kernel of each root of G', and apparently this could be done by methods similar to those given below. However, as the case of an irreducible admissible map is the most interesting one, (1.3) is probably enough.

For our main result, we shall make the assumptions which follow.

(i) G' is one of the simply-connected classical groups $Spin(m)$, $SU(n)$, $Sp(n)$.

(ii) $\theta: T \longrightarrow T'$ is an admissible map.

(iii) $\theta^*: R(G) \longleftarrow R(G')$ preserves real and symplectic elements.

(iv) $\theta\delta \in Z'$.

(v) If $G' = Spin(2n)$ with $2n \equiv 0$ mod 4, or if $G' = Sp(n)$, we assume that the map θ is irreducible in the sense of [2] p20.

To state our final assumption we need some notation. Let $x_1, x_2, \ldots, x_n$ be the basic weights in $G' = Spin(2n)$, $Spin(2n+1)$, $SU(n)$ or $Sp(n)$; and set $\theta_i = \theta^* x_i$, so that the map θ has components θ_i (except in the case $G' = SU(n)$, in which we have a relation $\sum_i \theta_i = 0$). Fix a Weyl chamber C in G, and let $\tau \in W$ be the element which carries C to -C.

(vi) If $G' = Spin(m)$ with $m \not\equiv 2 \bmod 4$ or $G' = SU(n)$ with $n \equiv 2 \bmod 4$ we assume that no θ_i is fixed by τ.

Theorem 1.4. Under these conditions we have $\theta\delta \equiv \delta' \bmod I'$.

We pause to comment on these assumptions. In (i), the only exceptional group we need to exclude is E_7, which would involve ad hoc calculations. We presume that for E_7 it would be appropriate to assume that θ is irreducible. Assumptions (ii) to (iv) seem acceptable. However assumption (iii) can sometimes be weakened, as we proceed to show. Lemma (1.2)(b) yields many cases in which $\delta' \in I'$, so that every self-conjugate representation of G' is real; in such cases θ automatically preserves symplectic elements, and so it is enough to assume that θ preserves real elements. The opposite happens when G' has enough basic symplectic representations. For example, if $G' = Sp(n)$ then it is sufficient to assume that $\theta^*\lambda^1$ is symplectic. For if so, then since θ^* commutes with λ^k, we see that $\theta^*\lambda^k$ is real or symplectic according to the parity of k, and similarly when λ^k is replaced by a monomial $(\lambda^1)^{e_1}(\lambda^2)^{e_2}\ldots(\lambda^n)^{e_n}$. Inspection of the proof of (1.4) will show what is actually used in each case.

Assumption (v) serves to rule out cases like the inclusion $SU(n) \subset Spin(2n)$ for $n \equiv 0 \bmod 2$ and the inclusion $SU(n) \subset Sp(n)$ for $n \equiv 1 \bmod 2$; in these cases the conclusion fails, although all the other assumptions hold. Assumption (vi) serves to rule out cases like the exterior powers $\lambda^{2i}: SU(n) \longrightarrow SU(m)$ where $m = n!/(2i)!(n-2i)!$ and

$m \equiv 2 \bmod 4$ (for example $\lambda^2 : SU(4) \longrightarrow SU(6)$). In these cases also the conclusion fails, although all the other assumptions hold. Assumption (vi) is not too restrictive; for many groups G we have $\tau = -1$, and then θ_i can be fixed by τ only if $\theta_i = 0$ - a possibility which is normally ruled out when θ is irreducible. In §5 we shall see how one can make use of assumption (vi).

The remainder of this paper is organised as follows. In §2 we prove Proposition 1.1. In §3 we deduce Lemma 1.2 from the statement originally given by Dynkin; we also prove Corollary 3.1, a useful result of representation-theory which however did not need to be stated in the introduction. In §4 we prove Proposition 1.3, and in §5 we prove the main result, Theorem 1.4.

The first author thanks the University of Kuwait for hospitality during the preparation of this paper.

§2. Preservation of real and symplectic elements. In this section we will prove Proposition 1.1.

We begin by studying the case in which the group G is simply-connected. In this case one of the main theorems of Hermann Weyl [1 p164] asserts that the representation ring R(G) is a polynomial algebra; and the proof of the theorem gives a preferred set of generators. More precisely, the weights in the closure of the fundamental dual Weyl chamber form a free (commutative) semigroup, with a unique set of generators, say $\omega_1, \omega_2, \ldots, \omega_\ell$ [1 p163]; to these weights there correspond irreducible representations $\rho_1, \rho_2, \ldots, \rho_\ell$; and the theorem says that R(G) is a polynomial algebra $Z[\rho_1, \rho_2, \ldots, \rho_\ell]$ on these generators.

The map $-1: L(T) \longrightarrow L(T)$ is conjugate to a map $-\tau$ (see §1) which preserves the fundamental Weyl chamber C. It follows that $-\tau$

permutes $\omega_1, \omega_2, \ldots, \omega_\ell$. Thus complex conjugation permutes the irreducible representations $\rho_1, \rho_2, \ldots, \rho_\ell$. Those of $\rho_1, \rho_2, \ldots, \rho_\ell$ which are self-conjugate are either real or symplectic, but not both [1 p64].

Complex conjugation also permutes the monomials $\rho_1^{i_1} \rho_2^{i_2} \ldots \rho_\ell^{i_\ell}$. Let us restrict attention to the self-conjugate monomials. We will call such a monomial "real" or "symplectic" according as the sum of the exponents of symplectic generators ρ_i is even or odd. Since a representation of the form $\rho\bar{\rho}$ is real [1 p166], it follows that each "real" monomial is a real representation, and similarly each "symplectic" monomial is a symplectic representation.

Lemma 2.1. The subgroup RO(G) of R(G) has a base consisting of the following elements.

(i) m, where m runs over the real monomials.

(ii) $m+\bar{m}$, where $(m,\bar{m})$ runs over pairs of distinct conjugate monomials.

(iii) 2m, where m runs over the symplectic monomials.

The subgroup RSp(G) of R(G) has a base consisting of the following elements.

(i) 2m, where m runs over the real monomials.

(ii) $m+\bar{m}$, where $(m,\bar{m})$ runs over pairs of distinct conjugate monomials.

(iii) m, where m runs over the symplectic monomials.

Proof. In view of the discussion preceding the lemma, it is clear that the subgroup generated by the elements listed is contained in RO(G) or RSp(G) as the case may be.

For the converse, we recall that it is easy to prove the corresponding descriptions of RO(G) and RSp(G) in terms of real irreducible representations, pairs of distinct complex conjugate irreducible representations, and symplectic irreducible representations; see for example [1 p66]. Any irreducible representation ρ can be written as a Z-linear combination of monomials, $\rho = \sum_I a_I m_I$, and therefore $\rho+\bar{\rho}$ can be written as $\rho+\bar{\rho} = \sum_I a_I(m_I+\bar{m}_I)$. Consider now a monomial $m = \rho_1^{i_1}\rho_2^{i_2}\ldots\rho_\ell^{i_\ell}$. Let σ be the irreducible representation corresponding to the "highest weight" $i_1\omega_1+i_2\omega_2+\ldots+i_\ell\omega_\ell$; then σ occurs in m with multiplicity 1 [1 pp161-164]. Thus m is self-conjugate or not with σ, and in the self-conjugate case, m is real or symplectic with σ. It now follows that when we write a self-conjugate irreducible representation σ' by induction as a Z-linear combination of monomials m', we use, apart from sums $m+\bar{m}$, only monomials m" which are real or symplectic with σ'. This proves the lemma.

We now introduce the reduced generators $\sigma_i = \rho_i-(\varepsilon\rho_i)1$, where ε is the augmentation. These generators lie in the augmentation ideal I(RG). We still have $R(G) = Z[\sigma_1,\sigma_2,\ldots,\sigma_\ell]$. Complex conjugation permutes $\sigma_1,\sigma_2,\ldots,\sigma_\ell$ just as it permuted $\rho_1,\rho_2,\ldots,\rho_\ell$; σ_i is self-conjugate or not with ρ_i; and in the self-conjugate case, σ_i is real or symplectic with ρ_i. We may form monomials $\sigma_1^{i_1}\sigma_2^{i_2}\ldots\sigma_\ell^{i_\ell}$; such a monomial is self-conjugate or not with $\rho_1^{i_1}\rho_2^{i_2}\ldots\rho_\ell^{i_\ell}$; and in the self-conjugate case, we call $\sigma_1^{i_1}\sigma_2^{i_2}\ldots\sigma_\ell^{i_\ell}$ "real" or "symplectic" with $\rho_1^{i_1}\rho_2^{i_2}\ldots\rho_\ell^{i_\ell}$.

Lemma 2.2. The description of RO(G) and RSp(G) in Lemma 2.1 remains valid if we use monomials $m = \sigma_1^{i_1}\sigma_2^{i_2}\ldots\sigma_\ell^{i_\ell}$ in the reduced generators.

This follows immediately from Lemma 2.1.

Next we recall that by theorems of Atiyah and Segal [3 pp10, 14,17], we have canonical isomorphisms

$$R(G)^{\wedge} \xrightarrow{\cong} K(BG)$$

$$RO(G)^{\wedge} \xrightarrow{\cong} KO(BG)$$

$$RSp(G)^{\wedge} \longrightarrow KSp(BG).$$

Here $R(G)^{\wedge}$ means the completion of $R(G)$ with respect to the topology defined by powers of the augmentation ideal $I(RG)$. The topologies on $RO(G)$, $RSp(G)$ may be taken to be the restrictions of the topology on $R(G)$; see [3 p17], but note that the reference there to (5.1) should be to (6.1). The following diagrams commute.

$$\begin{array}{ccc} RO(G)^{\wedge} & \xrightarrow{\cong} & KO(BG) \\ \downarrow & & \downarrow \\ R(G)^{\wedge} & \xrightarrow{\cong} & K(BG) \end{array}$$

$$\begin{array}{ccc} RSp(G)^{\wedge} & \xrightarrow{\cong} & KSp(BG) \\ \downarrow & & \downarrow \\ R(G)^{\wedge} & \xrightarrow{\cong} & K(BG) \end{array}$$

We proceed to describe these completions explicitly in our case.

Corollary 2.3. $R(G)^{\wedge}$ is the ring of formal power-series $Z[[\sigma_1,\sigma_2,\ldots,\sigma_\ell]]$. An element of $R(G)^{\wedge}$ may be written uniquely as a formal Z-linear combination $\sum_I a_I m_I$ $(a_I \in Z)$ of monomials $m_I = \sigma_1^{i_1}\sigma_2^{i_2}\ldots\sigma_\ell^{i_\ell}$. Such an element is self-conjugate if and only if conjugate monomials appear with equal coefficients. If self-conjugate, it lies in the completion $RO(G)^{\wedge}$ of $RO(G)$ if and only if the coefficient of each symplectic monomial is even; it lies in the completion $RSp(G)^{\wedge}$ of $RSp(G)$ if and only if the coefficient of each real monomial is even.

This follows immediately from the work above.

Corollary 2.4. In K(BG) we have

$$R(G) \cap KO(BG) = RO(G)$$

and

$$R(G) \cap KSp(BG) = RSp(G).$$

Proof. Take a typical element of $K(BG) = R(G)^{\wedge}$ as a formal Z-linear combination $\sum_I a_I m_I$. If it lies in R(G) only finitely many of the coefficients a_I are non-zero. If it lies in $KO(BG) = RO(G)^{\wedge}$ then conjugate monomials occur with equal coefficients, and the coefficient of each symplectic monomial is even. If the element lies both in R(G) and in $RO(G)^{\wedge}$, then it lies in RO(G). The proof of the second statement is similar.

Proof of Proposition 1.1. By Corollary 1.13 of [2], an induced map

$$f^*: K(BG) \longleftarrow K(BG')$$

carries R(G') into R(G); it clearly carries KO(BG') into KO(BG), and KSp(BG') into KSp(BG). So when G is simply-connected, Proposition 1.1 follows from Corollary 2.4.

If G is merely semi-simple, let $\tilde{G}$ be its universal cover, so that we have a finite covering map $\pi: \tilde{G} \longrightarrow G$. Then the preceding result applies to the composite

$$B\tilde{G} \xrightarrow{B\pi} BG \xrightarrow{f} BG'.$$

So from $x \in RO(G')$ we infer $(B\pi)^* f^* x \in RO(\tilde{G})$. From this it follows that $f^* x \in RO(G)$. If we assume $x \in RSp(G')$, a similar argument applies. This proves Proposition 1.1.

§3. Dynkin elements. We begin by recalling the work of Dynkin [5]; the reader may also consult Bourbaki [4]. These authors assume that G is semi-simple, so we make this assumption for the moment and remove it later. Then the construction given for δ is as follows. Choose a Weyl chamber C; this determines a base of simple roots $\phi_1, \phi_2, \ldots, \phi_\ell$. Let us consider these roots as linear maps $L(T) \longrightarrow R$. Then there is a unique vector $\tilde{t} \in L(T)$ such that $\phi_i(\tilde{t}) = 1$ for $i = 1, 2, \ldots, \ell$. It follows that every root takes an integer value on $\tilde{t}$; so $\tilde{t}$ maps to the identity under the adjoint representation, and $\tilde{t}$ yields an element δ of the centre.

The authors cited guarantee the behaviour of $\rho(\delta)$ in the case they consider; we will discuss the other assertions of Lemma 1.2.

The element δ constructed above appears to depend on the choice of C; however, any other Weyl chamber C' may be obtained from C by the action of an element $w \in W$, which will carry $\delta(C)$ onto $\delta(C')$. Since W fixes central elements, $\delta(C) = \delta(C')$, and so the construction is canonical.

In particular, -C is another Weyl chamber; it leads to simple roots $-\phi_1, -\phi_2, \ldots, -\phi_\ell$ and to the vector $-\tilde{t}$. Hence $\delta^{-1} = \delta$ and $\delta^2 = 1$.

This establishes assertions (i) and (ii) of the lemma when G is semi-simple. Moreover, the construction of δ yields the following further properties. (iii) If G is a product group $G_1 \times G_2$, then the Dynkin element δ in G is the product of the Dynkin elements δ_1, δ_2 in G_1, G_2. (iv) Let $\tilde{G} \longrightarrow G$ be a finite covering; then the Dynkin element $\tilde{\delta}$ in $\tilde{G}$ maps to the Dynkin element δ in G. If we insist on preserving these two properties, and define the Dynkin element of any torus to be 1, we get a canonical extension of δ from the class of compact connected semi-simple groups to the class of compact connected groups; and this extension has the properties (i),(ii) stated in the lemma.

Part (b) lists the value of δ in each simple Lie group, and this is an easy calculation.

It remains to prove part (c), about $\delta\zeta^2$. Since $\delta^2 = 1$ and we assume $\zeta^4 = 1$, and since δ and ζ are both central, it is clear that $(\delta\zeta^2)^2 = 1$. Suppose then that ρ is an irreducible self-conjugate representation of G, as in the lemma. Since ρ is self-conjugate we get $\overline{\rho(\zeta)} = \rho(\zeta)$, that is $\rho(\zeta)^{-1} = \rho(\zeta)$, so $\rho(\zeta^2) = 1$ and $\rho(\delta\zeta^2) = \rho(\delta)$. This completes the proof of Lemma 1.2.

Next we recall that weights ω may be considered as 1-dimensional representations of T, and symmetric sums $S(\omega)$ as representations of T. We note that a central element $z \in Z$ acts as a scalar under $S(\omega)$; in fact, since W fixes central elements we get $(\omega w)z = \omega(wz) = \omega(z)$ for each w.

We shall call a self-conjugate symmetric sum $S(\omega)$ "real" or "symplectic" according as $S(\omega)\delta$ is +1 or -1.

<u>Corollary 3.1</u>. The subgroup RO(G) of R(G) has a base consisting of the following elements.

(i) $S(\omega)$ where $S(\omega)$ runs over the real symmetric sums.

(ii) $S(\omega)+\overline{S(\omega)}$, where $(S(\omega),\overline{S(\omega)})$ runs over pairs of distinct conjugates.

(iii) $2S(\omega)$, where $S(\omega)$ runs over the symplectic symmetric sums.

The subgroup RSp(G) of R(G) has a base consisting of the following elements.

(i) $2S(\omega)$, where $S(\omega)$ runs over the real symmetric sums.

(ii) $S(\omega)+\overline{S(\omega)}$, where $(S(\omega),\overline{S(\omega)})$ runs over pairs of distinct conjugates.

(iii) $S(\omega)$ where $S(\omega)$ runs over the symplectic symmetric sums.

If we were allowed to replace the symmetric sum $S(\omega)$ by the irreducible representation $\rho(\omega)$ with ω as an extreme weight, this would become a standard result of representation-theory; see [1] p66. We can write each $\rho(\omega)$ in terms of the $S(\omega')$, and (by induction over ω) each $S(\omega)$ in terms of the $\rho(\omega')$; the work above assures us that in this process we only use weights ω' with a fixed value of $\omega'(\delta)$.

§4. Preservation of the centre. We interpret the Stiefel diagram $L(T)$ as the universal cover of T. As in [2 p22], let Γ be the extended Weyl group of G (generated by the reflections in all the planes of the Stiefel diagram, not just those which pass through the origin); and let Γ_o be the subgroup of translations in Γ.

Lemma 4.1. Let $f: BG \longrightarrow BG'$ correspond to $\theta: T \longrightarrow T'$; then $\tilde{\theta} = L(\theta)$ carries Γ_o into Γ_o'.

Proof. f induces a map from $\pi_1(G) = \pi_2(BG)$ to $\pi_1(G') = \pi_2(BG')$. Here we can suppose without loss of generality that $\pi_1(G')$ is free abelian, for we can arrange this by passing to finite covers of G and G' - a step which does not change $\tilde{\theta} = L(\theta)$. The diagram

$$\begin{array}{ccc} \pi_1(T) = \pi_2(BT) & \xrightarrow{(B\theta)_*} & \pi_2(BT') = \pi_1(T') \\ \downarrow \qquad \downarrow & & \downarrow \qquad \downarrow \\ \pi_1(G) = \pi_2(BG) & \xrightarrow{f_*} & \pi_2(BG') = \pi_1(G') \end{array}$$

is now commutative, for when $\pi_1(G')$ is free abelian this follows from the results on rational cohomology in [2]. We can interpret Γ_o as the kernel of $\pi_1(T) \longrightarrow \pi_1(G)$, and Γ_o' as the kernel of $\pi_1(T') \longrightarrow \pi_1(G')$; hence $\tilde{\theta} = L(\theta)$ carries Γ_o into Γ_o'.

Proof of Proposition 1.3. Take an element $z \in Z$, and lift it to a vector $\tilde{z} \in \tilde{T} = L(T)$. By Proposition 2.28 of [2] (applied with G' replaced by G) we have

$$w\tilde{z} \equiv \tilde{z} \bmod \Gamma_o$$

for all $w \in W$. Let $\alpha: W \longrightarrow W'$ be as in Theorem 2.29 of [2]; applying $\tilde{\theta} = L(\theta)$ and using Lemma 4.1, we get

$$(\alpha w)(\tilde{\theta}\tilde{z}) = (\tilde{\theta}\tilde{z}) \bmod \Gamma'_o.$$

Since in Proposition 1.3 we assume θ irreducible, Theorem 2.29(ii) of [2] yields

$$\tilde{\theta}\tilde{z} \in \tilde{Z}'.$$

So θ carries Z into Z'.

§5. Location of self-conjugate symmetric sums. The pattern of proof of Theorem 1.4 is as follows. We first select some irreducible self-conjugate representations ρ'_i of G'. We take care to choose enough representations ρ'_i so that the conditions

$$z' \in Z', \quad (z')^2 = 1 \text{ and } \rho'_i(z') = 1 \text{ for all } i$$

imply $z' \in I'$; we shall see that this can be done in each case. It is then sufficient to prove that

$$\rho'_i(\theta\delta) = \rho'_i(\delta') \text{ for each } i.$$

Since we assume that θ^* preserves real and symplectic elements, $\theta^*\rho'_i$ is real or symplectic with ρ'_i. The crucial step is now to prove that $\theta^*\rho'_i$ contains at least one self-conjugate symmetric sum $S(\omega_i)$ with odd multiplicity. If so, then we can apply Corollary 3.1 to the expression for $\theta^*\rho'_i$ in terms of symmetric sums, and conclude that $S(\omega_i)$ is real or symplectic with $\theta^*\rho'_i$ and ρ'_i; that is,

$$S(\omega_i)(\delta) = \rho'_i(\delta').$$

But since $S(\omega_i)$ occurs in $\theta^*\rho'_i$, we also have

$$\rho'_i(\theta\delta) = S(\omega_i)\delta.$$

Thus

$$\rho_i'(\theta\delta) = \rho_i'(\delta')$$

and this completes the proof.

To fill in this outline, we must first show that we can find enough representations ρ_i'. Let us dismiss as trivial the cases

$G' = \mathrm{Spin}(n)$ with $n \equiv 2 \bmod 4$ and

$G' = SU(n)$ with $n \not\equiv 2 \bmod 4$.

In these cases the conditions

$$z' \in Z', \ (z')^2 = 1$$

already imply $z' \in I'$, so we need choose no representations ρ_i'.

Next we take the cases

$\mathrm{Spin}(n)$ with $n \equiv 1 \bmod 2$

$SU(n)$ with $n \equiv 2 \bmod 4$ and

$Sp(n)$.

In each case we have just 2 elements $z' \in Z'$ such that $(z')^2 = 1$, and one representation ρ_i' will do. It is sufficient to take the spin-representation Δ on $\mathrm{Spin}(n)$ with $n \equiv 1 \bmod 2$, the exterior power λ^m on $SU(2m)$ where $2m \equiv 2 \bmod 4$, and the fundamental representation λ^1 on $Sp(n)$.

Finally we take the case

$\mathrm{Spin}(n)$ with $n \equiv 0 \bmod 4$.

In this case the centre is $Z_2 \times Z_2$. It is sufficient to take two representations ρ_i', namely the fundamental representation λ^1, and either one of the two half-spin representations Δ^+, Δ^-. This completes the choice of the ρ_i'.

The essential step is now to show in each case that $\theta^*\rho_i'$ contains at least one self-conjugate symmetric sum $S(\omega_i)$ with odd multiplicity. We consider first the cases

$$\mathrm{Spin}(2n) \text{ with } 2n \equiv 0 \bmod 4$$
$$\mathrm{Sp}(n)$$

in which we have to consider $\rho_i' = \lambda^1$. In both cases W' permutes the 2n weights $\pm x_1, \pm x_2, \ldots, \pm x_n$; therefore W permutes the 2n elements $\pm\theta_1, \pm\theta_2, \ldots, \pm\theta_n$, and they fall into orbits under W. If $(\phi_1, \phi_2, \ldots, \phi_r)$ is an orbit, then $(-\phi_1, -\phi_2, \ldots, -\phi_r)$ is also an orbit. Suppose to begin with that one of the orbits other than $(\phi_1, \phi_2, \ldots, \phi_r)$ has the same elements as $(-\phi_1, -\phi_2, \ldots, -\phi_r)$. Then we can factor the admissible map θ; if $G' = \mathrm{Sp}(n)$ we use the subgroup $U(r)\times\mathrm{Sp}(n-r)$, while if $G' = \mathrm{Spin}(2n)$ we use the pull back in the following diagram.

$$\begin{array}{ccccc} G'' & & \longrightarrow & & \mathrm{Spin}(2n) \\ \downarrow & & & & \downarrow \\ U(r)\times SO(2n-2r) & \longrightarrow & SO(2r)\times SO(2n-2r) & \longrightarrow & SO(2n) \end{array}$$

This contradicts the assumption that θ is irreducible; therefore, it does not happen. We conclude (firstly) that $(-\phi_1, -\phi_2, \ldots, -\phi_r)$ is the same orbit as $(\phi_1, \phi_2, \ldots, \phi_r)$; therefore, each orbit has the form $(\pm\psi_1, \pm\psi_2, \ldots, \pm\psi_s)$. We also conclude that no orbit is repeated, for if the orbit $(\phi_1, \phi_2, \ldots, \phi_r)$ were repeated, the second copy of $(\phi_1, \phi_2, \ldots, \phi_r)$ would have the same elements as $(-\phi_1, -\phi_2, \ldots, -\phi_r)$. This says that when we decompose $\theta^*\lambda^1$ into symmetric sums, corresponding to the orbits $(\pm\psi_1, \pm\psi_2, \ldots, \pm\psi_s)$, each symmetric sum is self-conjugate and occurs with multiplicity 1.

In the remaining cases we have to consider $\rho_i' = \Delta$, λ^m and either Δ^+ or Δ^-. We use the same argument in all cases. Let

$\phi_1, \phi_2, \ldots, \phi_{2m}$ be

(i) $\pm\frac{1}{2}\theta_1, \pm\frac{1}{2}\theta_2, \ldots, \pm\frac{1}{2}\theta_m$ in the cases $G' = \mathrm{Spin}(2m)$ and $G' = \mathrm{Spin}(2m+1)$

(ii) $\theta_1, \theta_2, \ldots, \theta_{2m}$ in the case $G' = SU(2m)$.

In either case we have

$$\sum_1^{2m} \phi_j = 0.$$

By assumption (vi), ϕ_j is not fixed by τ, so the condition $\phi_j(\tau-1)v \neq 0$ is satisfied by an open dense set of vectors $v \in L(T)$. It follows that we can find a vector v such that $\phi_j(\tau-1)v \neq 0$ for all j. Now, τ permutes $\phi_1, \phi_2, \ldots, \phi_{2m}$, and for the element $\phi_j\tau$ we have

$$\begin{aligned}\phi_j\tau(\tau-1)v &= \phi_j(1-\tau)v \\ &= -\phi_j(\tau-1)v.\end{aligned}$$

So the elements $\phi_1, \phi_2, \ldots, \phi_{2n}$ fall into pairs $(\phi, \phi\tau)$, with one member of each pair being positive on $(\tau-1)v$ and one member of each pair being negative. Without loss of generality, we can suppose the ϕ's renumbered so that $\phi_1, \phi_2, \ldots, \phi_m$ are positive on $(\tau-1)v$. We claim that $\omega_i = \phi_1+\phi_2+\ldots+\phi_m$ is one of the weights of $\theta^*\rho_i'$. This is clear in the case $G' = SU(2m)$, $\rho_i' = \lambda^m$ since ω_i is merely the sum of some m of the θ_i. In the cases $G' = \mathrm{Spin}(2m)$ and $\mathrm{Spin}(2m+1)$ the ϕ's already fall into pairs

$$\pm\frac{1}{2}\theta_1, \pm\frac{1}{2}\theta_2, \ldots, \pm\frac{1}{2}\theta_m$$

taking opposite values on $(\tau-1)v$; we must have selected one from each pair and obtained

$$\omega_i = \frac{1}{2}(\varepsilon_1\theta_1+\varepsilon_2\theta_2+\ldots+\varepsilon_m\theta_m)$$

where each ε_i is ± 1. If $G = \mathrm{Spin}(2m+1)$ this is one of the weights of

$\theta^*\Delta$, because $\frac{1}{2}(\varepsilon_1 x_1+\varepsilon_2 x_2+\ldots+\varepsilon_m x_m)$ is a weight of Δ. If $G = \mathrm{Spin}(2m)$ then ω_i is either one of the weights of $\theta^*\Delta^+$ or one of the weights of $\theta^*\Delta^-$.

Next we claim that this weight occurs with multiplicity 1 in $\theta^*\rho_i'$. In fact, among the weights of $\theta^*\rho_i'$, ω_i is by construction the one with the maximum value at $(\tau-1)v$.

Finally we claim that the symmetric sum $S(\omega_i)$ is self-conjugate. In fact, by construction we have

$$\begin{aligned}\omega_i\tau &= \phi_1\tau+\phi_2\tau+\ldots+\phi_m\tau \\ &= \phi_{m+1}+\phi_{m+2}+\ldots+\phi_{2m} \\ &= -(\phi_1+\phi_2+\ldots+\phi_m) \quad (\text{since } \sum_1^{2m}\phi_j = 0) \\ &= -\omega_i.\end{aligned}$$

This proves the required result for $\theta^*\Delta$ and for $\theta^*\lambda^m$; and for $G' = \mathrm{Spin}(n)$ with $n \equiv 0 \bmod 4$ it shows that the required result holds either for $\theta^*\Delta^+$ or for $\theta^*\Delta^-$. This is enough, and it completes the proof.

References

[1] J.F. Adams, "Lectures on Lie Groups", W.A. Benjamin 1969; to be reprinted by the University of Chicago Press.

[2] J.F. Adams and Z. Mahmud, "Maps between classifying spaces", Inventiones Math. 35 (1976) 1-41.

[3] M.F. Atiyah and G.B. Segal, "Equivariant K-theory and completion", Jour. Differential Geometry 3 (1969) 1-18.

[4] N. Bourbaki, "Groups et algebras de Lie" Chap.VIII (especially pages 131-133), Hermann, Paris 1975.

[5] E.B. Dynkin, "Maximal subgroups of the classical groups", in American Math. Soc. translations, series 2, volume 6, American Math. Soc. 1957, 245-378.

[6] J. Hubbuck, "Homotopy homomorphisms of Lie groups", in "New developments in topology", L.M.S. Lecture Note Series no.11, C.U.P. 1974, 33-41, especially pages 33-34.

References (cont.)

[7] Z. Mahmud, "The maps BSp(1) ⟶ BSp(n)", Proc. Amer. Math. Soc. 52 (1975) 473-478.

IDEMPOTENT CODENSITY MONADS AND THE PROFINITE COMPLETION OF TOPOLOGICAL GROUPS

A. Deleanu
Syracuse University, Syracuse, New York, U.S.A.

The profinite completion defined on the category of groups [11], [12] presents the difficulty that it is not idempotent, that is, the iterated profinite completion is not equivalent to the single one [1], [8]. By developing an idea of Frank Adams [1], [2], it is shown in this paper that this difficulty can be avoided by defining profinite completion as a functor on the category of <u>topological</u> groups.

The framework of codensity monads is used in order to describe various profinite completion functors, and a general criterion for such a monad to be idempotent is established.

Idempotent monads were discussed by Peter Hilton, Armin Frei and the author in [7].

1 Idempotent codensity monads

Recall that a <u>monad</u> (or <u>triple</u>) $<T, \eta, \mu>$ on a category $\underline{C}$ [9] consists of a functor $T : \underline{C} \to \underline{C}$ and two natural transformations $\eta : 1_{\underline{C}} \to T$, $\mu : T^2 \to T$ which make the following diagrams commute:

$$\begin{array}{ccc} T^3 & \xrightarrow{T\mu} & T^2 \\ {\scriptstyle \mu T}\downarrow & & \downarrow{\scriptstyle \mu} \\ T^2 & \xrightarrow{\mu} & T \end{array} , \qquad \begin{array}{ccccc} 1_{\underline{C}}T & \xrightarrow{\eta T} & T^2 & \xleftarrow{T\eta} & T1_{\underline{C}} \\ \| & & {\scriptstyle \mu}\downarrow & & \| \\ T & = & T & = & T \end{array} .$$

A monad $<T, \eta, \mu>$ is called <u>idempotent</u> if $\eta_{TX} : TX \to T^2X$ is an isomorphism for all $X \in \underline{C}$. Plainly, $<T, \eta, \mu>$ is idempotent if and only if μ is a natural equivalence.

<u>PROPOSITION 1.1</u>. <u>The monad</u> $<T, \eta, \mu>$ <u>is idempotent if and only if</u> $\eta T = T\eta$.

Proof. Assume $<T, \eta, \mu>$ is idempotent. Since $\mu \cdot \eta T = 1_T = \mu \cdot T\eta$ and μ is a natural equivalence, we have $\eta T = \mu^{-1} = T\eta$. Conversely, suppose $\eta T = T\eta$. The naturality of η yields the commutative diagram

$$\begin{array}{ccc} T^2 & \xrightarrow{\eta T^2} & T^3 \\ \mu \downarrow & & \downarrow T\mu \\ T & \xrightarrow{\eta T} & T^2 \end{array}$$

But from our assumption we infer $\eta T^2 = T\eta T$, so that

$$\eta T \cdot \mu = T\mu \cdot \eta T^2 = T\mu \cdot T\eta T = T(\mu \cdot \eta T) = T(1_T) = 1_{T^2};$$

this, combined with $\mu \cdot \eta T = 1_T$, shows that μ is an equivalence.

Now, let $K : \underline{M} \to \underline{C}$ be a functor such that the right Kan extension $R : \underline{C} \to \underline{C}$ of K along K exists. Thus [9], we have a natural transformation $\varepsilon : RK \to K$ such that, for any functor $S : \underline{C} \to \underline{C}$, the assignment $\sigma \mapsto \varepsilon \cdot \sigma K$ is a bijection

$$\varphi : \mathrm{Nat}(S,R) \cong \mathrm{Nat}(SK,K).$$

We set $S = 1_{\underline{C}}$ and define $\eta : 1_{\underline{C}} \to R$ by $\eta = \varphi^{-1}(1_K)$; we then set $S = R^2$ and define $\mu : R^2 \to R$ by $\mu = \varphi^{-1}(\varepsilon \cdot R\varepsilon)$. It is easy to check that η and μ are natural, and that $\langle R, \eta, \mu \rangle$ is a monad on $\underline{C}$; in fact, this is called the codensity monad of K in [9, p.246].

From now on, we shall assume that the pointwise right Kan extension of K along K exists. Recall that this is described in terms of a limit formula, as follows [9]: For each $X \in \underline{C}$, the comma category $(X \downarrow K)$ has as objects the maps $f : X \to KZ$ in $\underline{C}$, where $Z \in \underline{M}$, and as morphisms from f to $f' : X \to KZ'$ those maps $\ell : Z \to Z'$ in $\underline{M}$ for which $(K\ell)f = f'$; there is a projection functor $Q_X : (X \downarrow K) \to \underline{M}$ sending f to Z. Then $RX = \varprojlim KQ_X$, and, if $h : X \to Y$ in $\underline{C}$, Rh is the unique morphism in $\underline{C}$ which makes the diagram

$$\begin{array}{ccc} RX = \varprojlim KQ_X & \xrightarrow{\lambda^X_{gh}} & KZ \\ Rh \downarrow & & \| \\ RY = \varprojlim KQ_Y & \xrightarrow{\lambda^Y_g} & KZ \end{array}$$

commute for every $g : Y \to KZ$ in $(Y{\downarrow}K)$, where λ^X (resp. λ^Y) is the limiting (universal) cone of $\varprojlim KQ_X$ (resp. $\varprojlim KQ_Y$).

The description of η in terms of limits is given by

PROPOSITION 1.2. *For each* $X \in \underline{C}$, *let* σ_X *be the unique morphism in* $\underline{C}$ *which makes the diagram*

$$\begin{array}{ccc} & \overset{\sigma_X}{\nearrow} & RX = \varprojlim KQ_X \\ X & & \downarrow \lambda^X_f \\ & \underset{f}{\searrow} & KZ = KQ_X(f) \end{array} \qquad (1)$$

commute for every $f \in (X{\downarrow}K)$. *Then* $\sigma : 1_K \to R$ *is a natural transformation and* $\sigma = \eta$.

Proof. To check the naturality of σ, let $h : X \to Y$ in $\underline{C}$. Then, in view of the above description of Rh, we can write for every $g \in (Y{\downarrow}K)$

$$\lambda^Y_g(Rh)\sigma_X = \lambda^X_{gh}\sigma_X = gh = \lambda^Y_g\sigma_Y h,$$

which implies $(Rh)\sigma_X = \sigma_Y h$.

To prove that $\sigma = \eta$, we must show that $\varphi(\sigma) = 1_K$, that is $\epsilon \cdot \sigma K = 1_K$. Taking $X = KZ$ and $f = 1_{KZ}$ in diagram (1), we get $\lambda^{KZ}_{1_{KZ}}\sigma_{KZ} = 1_{KZ}$. But, by [9, p.234], $\lambda^{KZ}_{1_{KZ}} = \epsilon_Z$.

We shall need in the sequel the following

DEFINITION [10, p.118]. Let $u : A \to B$ be a morphism in a category $\underline{D}$ and let C be an object of $\underline{D}$. u is said to be *epimorphic relative to* C if

$$\underline{D}(u,C) : \underline{D}(B,C) \to \underline{D}(A,C)$$

is an injective map.

THEOREM 1.3. *The codensity monad* $\langle R, \eta, \mu \rangle$ *of* $K : \underline{M} \to \underline{C}$ *is idempotent if and only if, for each* $X \in \underline{C}$, η_X *is epimorphic relative to* KZ *for every* $Z \in \underline{M}$.

Proof. First, note that for each $X \in \underline{C}$, by Proposition 1.2, η_{RX} is determined by the condition that the diagram

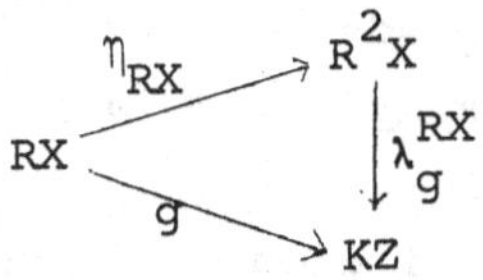

$$\begin{array}{ccc} RX & \xrightarrow{\eta_{RX}} & R^2X \\ & {\scriptstyle g}\searrow & \downarrow{\scriptstyle \lambda_g^{RX}} \\ & & KZ \end{array} \tag{2}$$

commutes for every $g \in (RX{\downarrow}K)$, whereas, by the above description in terms of limits, $R(\eta_X)$ is determined by the condition that the diagram

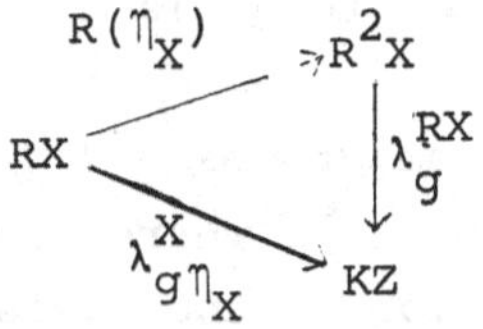

$$\begin{array}{ccc} RX & \xrightarrow{R(\eta_X)} & R^2X \\ & {\scriptstyle \lambda_{g\eta_X}^X}\searrow & \downarrow{\scriptstyle \lambda_g^{RX}} \\ & & KZ \end{array} \tag{3}$$

commutes for every $g \in (RX{\downarrow}K)$.

Now, assume that, for each $X \in \underline{C}$, η_X is epimorphic relative to KZ for every $Z \in \underline{M}$. This implies, since by Proposition 1.2 there is for each $g \in (RX{\downarrow}K)$ a commutative diagram

$$\begin{array}{ccc} X & \xrightarrow{\eta_X} & RX \\ & {\scriptstyle g\eta_X}\searrow & \downarrow{\scriptstyle \lambda_{g\eta_X}^X} \\ & & KZ \end{array},$$

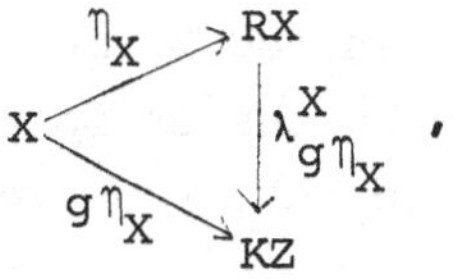

that $\lambda_{g\eta_X}^X = g$. Then diagrams (2) and (3) show that $\eta_{RX} = R(\eta_X)$, so that, by Proposition 1.1, $\langle R, \eta, \mu \rangle$ is idempotent.

Conversely, suppose that $\langle R, \eta, \mu \rangle$ is idempotent. Then, by Proposition 1.1, $\eta_{RX} = R(\eta_X)$ for each $X \in \underline{C}$. This implies, in view of diagrams (2) and (3), that $\lambda_{g\eta_X}^X = g$ for each $g \in (RX{\downarrow}K)$. Thus, if two morphisms

$$RX \underset{\ell}{\overset{h}{\rightrightarrows}} KZ$$

are given such that $h\eta_X = \ell\eta_X$, we have

$$h = \lambda_{h\eta_X}^X = \lambda_{\ell\eta_X}^X = \ell,$$

so that η_X is epimorphic relative to KZ for each $Z \in \underline{M}$.

2 Examples of codensity monads

Throughout the following examples, K will be the appropriate inclusion functor.

(1) Let C be the category of groups and M the full subcategory of finite groups. Then R is the profinite completion of groups [11], [12]. This monad is not idempotent, as shown by the example of a direct sum of an infinite number of cyclic groups of order p, where p is a fixed prime [1], [8].

(2) Let C be the category of groups and M the full subcategory of A-nilpotent groups [5], where A is a solid ring. Then R is the A-completion in the sense of Bousfield and Kan [5, p.103]. This monad is not idempotent, as shown by the example of a free group on a countably infinite set of generators, with $A = \mathbb{Z}$ [4, p.57].

(3) Let C be the homotopy category of pointed connected CW-complexes and M the full subcategory of those which have finite homotopy groups. Then R is the Sullivan profinite completion [1], [11], [12], denoted by Su in [6]. This monad is not idempotent, as shown by the example of an Eilenberg-MacLane space corresponding to the group mentioned in example (1) above [1].

(4) Let C be the category of topologized objects of the homotopy category of pointed connected CW-complexes, as defined in [1], [2], [6], and M the full subcategory of those topologized objects whose underlying CW-complexes have finite homotopy groups. Then R is a generalized profinite completion, denoted by Su^T in [6]. The question as to whether this monad is idempotent is open.

(5) Let C be the category of topological groups and M the full subcategory of finite groups with discrete topology. Then R is the profinite completion of topological groups. This monad is idempotent, as shown in Theorem 3.1 below.

REMARK. More generally, throughout the preceding examples, we can replace "finite groups" by "finite P-groups, where P is an arbitrary family of primes." We thus obtain P-profinite completion functors.

3 The profinite completion of topological groups

THEOREM 3.1. *Let $\underline{C}$ be the category of topological groups and continuous homomorphisms, and let $\underline{M}$ be the full subcategory whose objects are the finite groups with the discrete topology. Then the codensity monad* $\langle R, \eta, \mu \rangle$ *of the inclusion functor* $K : \underline{M} \to \underline{C}$ *is idempotent.*

Proof. First, for each topological group G, consider the family $\{N_\alpha\}$ of all closed normal subgroups of G of finite index, and denote by $\underline{J}$ the subcategory of $(G{\downarrow}K)$ whose objects are the canonical projections onto quotient groups

$$p_\alpha : G \to K(G/N_\alpha) ,$$

and whose morphisms are the canonical homomorphisms

$$q_{\alpha\beta} : G/N_\alpha \to G/N_\beta \quad \text{whenever} \quad N_\alpha \subset N_\beta .$$

Each such subgroup is also open, so that G/N_α is indeed a discrete finite group.

We show that the inclusion functor $I : \underline{J} \to (G{\downarrow}K)$ is *initial* [9, p.214].

First, given an arbitrary object of $(G{\downarrow}K)$, $f : G \to KF$, there is a commutative diagram in $\underline{C}$

$$\begin{array}{ccc} G & \xrightarrow{p} & K(G/\ker f) \\ & \searrow^{f} & \downarrow{\scriptstyle Ks} \\ & & KF \end{array} ,$$

and $\ker f$ is plainly a closed normal subgroup of G of finite index.

Second, given a commutative diagram in $\underline{C}$

$$\begin{array}{ccccc} & \nearrow^{p_\alpha} & K(G/N_\alpha) & \searrow^{Ks} & \\ G & & \xrightarrow{\quad f \quad} & & KF \\ & \searrow_{p_\beta} & K(G/N_\beta) & \nearrow_{Kt} & \end{array} ,$$

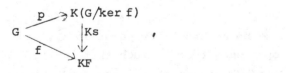

$N_\alpha \cap N_\beta$ is a closed normal subgroup of G of finite index, so that $N_\alpha \cap N_\beta = N_\gamma$ for some index γ, and we have the commutative diagram in $\underline{C}$

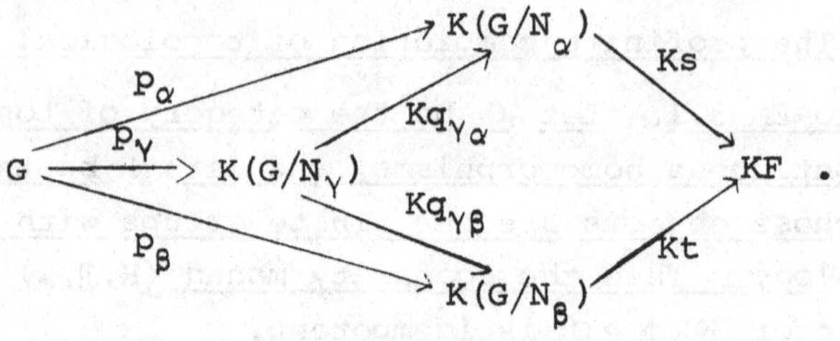

Now, since I is initial and $\varprojlim KQ_G I$ obviously exists, by the dual of Theorem 1 on p.213 of [9] $\varprojlim KQ_G$ also exists and there is a canonical isomorphism

$$RG = \varprojlim KQ_G \xrightarrow[\cong]{\omega_G} \varprojlim KQ_G I = \varprojlim_{\alpha} K(G/N_\alpha)$$

such that $\nu_{p_\alpha} \omega_G = \lambda^G_{p_\alpha}$ for each $p_\alpha \in \underline{J}$, where ν is the limiting cone of $\varprojlim KQ_G I$.

If we put $\tilde{G} = \varprojlim_{\alpha} K(G/N_\alpha)$ and $\theta_G = \omega_G \eta_G$, then, for each topological group G,

$$\theta_G : G \to \tilde{G}$$

is the unique map such that $\nu_{p_\alpha} \theta_G = p_\alpha$ for each $p_\alpha : G \to K(G/N_\alpha)$. For, by Proposition 1.2,

$$\nu_{p_\alpha} \theta_G = \nu_{p_\alpha} \omega_G \eta_G = \lambda^G_{p_\alpha} \eta_G = p_\alpha .$$

Next, we show that the image of θ_G is <u>dense</u> in $\tilde{G}$. To this end, let x be an arbitrary point in $\tilde{G}$, and let 0 be an arbitrary open subset of $\tilde{G}$ such that $x \in 0$. Then $x = (x_\alpha)$, where $x_\alpha \in G/N_\alpha$ for each α, and $0 = V \cap \tilde{G}$, where V is an open subset of $\prod_\alpha (G/N_\alpha)$. Thus $x \in \prod_\alpha U_\alpha \subset V$, where U_α is an open subset of G/N_α for each α and $U_\alpha = G/N_\alpha$ for all but a finite set of indices, say $\alpha_1, \ldots, \alpha_n$. Now $\bigcap_{i=1}^{n} N_{\alpha_i}$ is a closed normal subgroup of G of finite index, so that $\bigcap_{i=1}^{n} N_{\alpha_i} = N_\beta$ for some index β. Let $y \in G$ be such that $p_\beta(y) = x_\beta$. Then $p_\alpha(y) = q_{\beta\alpha} p_\beta(y) = x_\alpha$ for $\alpha = \alpha_1, \ldots, \alpha_n$. Thus $\nu_{p_\alpha} \theta_G(y) = p_\alpha(y) = x_\alpha$ for $\alpha = \alpha_1, \ldots, \alpha_n$, so that

$\theta_G(y) \in \prod_\alpha U_\alpha$. Hence $\theta_G(y) \in 0$, and x belongs to the closure of the image of θ_G.

The fact that the image of θ_G is dense in $\tilde{G}$ implies that θ_G, and therefore also η_G, is epimorphic relative to any Hausdorff topological group, in particular relative to KF for every discrete finite group F. But according to Theorem 1.3 this means that $\langle R, \eta, \mu \rangle$ is idempotent, which concludes the proof of the theorem.

Thus, for every topological group G, we have the <u>profinite completion</u> of G:

$$RG \cong \tilde{G} = \varprojlim_{\alpha} K(G/N_\alpha) .$$

We can describe this completion in terms of completions of uniform spaces, in a manner which is similar to the p-adic completion of the additive group of integers, as follows:

Take the family $\{N_\alpha\}$ of closed normal subgroups of G of finite index as a fundamental system of neighbourhoods of the identity element, thereby defining a second topology on the group underlying G, which is compatible with the group structure; let G' denote this new topological group. Being closed and of finite index, each N_α is also open in G, so that the homomorphism $i : G \to G'$ defined by $i(x) = x$ for all $x \in G$ is continuous. This also implies that $\{N_\alpha\}$ coincides with the family of closed normal subgroups of G' of finite index.

By considering the uniform space structure associated to G', we can form the <u>Hausdorff completion</u> $\widehat{G'}$ of G', which comes equipped with a canonical continuous homomorphism $\psi : G' \to \widehat{G'}$ [3, p.248]. The existence of $\widehat{G'}$ is guaranteed by Corollary 2 on p.290 of [3], which also yields the existence of an isomorphism τ' in <u>C</u> such that the diagram

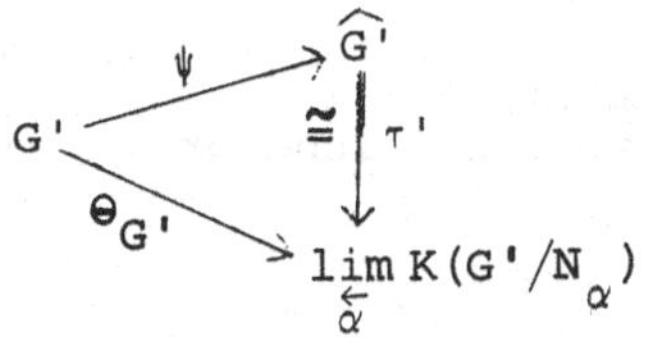

commutes.

But each N_α is open in both topological groups G and G', so that both G/N_α and G'/N_α are discrete. Consequently $\varprojlim_\alpha K(G/N_\alpha)$ and $\varprojlim_\alpha K(G'/N_\alpha)$ are isomorphic <u>qua</u> topological groups, and we conclude that there exists an isomorphism τ in <u>C</u> such that the diagram

$$\begin{array}{ccc} & \overset{\psi i}{\nearrow} & \hat{G}' \\ G & & \cong \Big\downarrow \tau \\ & \underset{\theta_G}{\searrow} & \tilde{G} \end{array}$$

commutes.

<u>Remark</u>. The author is very grateful to the referee for pointing out the papers of Lambek and Rattray [14], [15], [16], [17]; their work starts with a codensity monad on a complete category and then uses a method due to Fakir [13] in order to turn it into an idempotent monad. They give extensive applications to localizations and completions.

<u>References</u>

[1] Adams, J.F. (1975). Localization and completion. Lecture Notes in Mathematics, University of Chicago.

[2] Adams, J.F. (1975). Adams' Problems. <u>In</u> Manifolds-Tokyo 1973, University of Tokyo Press, Tokyo, pp. 430-431.

[3] Bourbaki, N. (1966). Elements of Mathematics. General Topology, Part 1, Hermann, Paris.

[4] Bousfield, A.K. (1977). Homological localization towers for groups and Π-modules. Memoirs of the Amer. Math. Soc. <u>10</u>, no. 186.

[5] Bousfield, A.K. and Kan, D.M. (1972). Homotopy limits, completions, and localizations. Lecture Notes in Mathematics No. 304, Springer-Verlag, Berlin.

[6] Deleanu, A. (1982). Topologized objects in categories and the Sullivan profinite completion. Journal of Pure and Applied Algebra <u>25</u>, pp. 21-24.

[7] Deleanu, A., Frei, A. and Hilton, P. (1975). Idempotent triples and completion. Math. Zeitschrift <u>143</u>, pp. 91-104.

[8] Huber, M. and Warfield, R.B. (1982). p-adic and p-cotorsion completions of nilpotent groups. Journal of Algebra <u>74</u>, pp. 402-442.

[9] Mac Lane, S. (1971). Categories for the working mathematician. Springer-Verlag, Berlin.

[10] Pareigis, B. (1970). Categories and functors. Academic Press, New York.

[11] Sullivan, D. (June 1970, revised April 1971). Geometric topology, Part I: localization, periodicity and Galois symmetry. M.I.T. Notes.
[12] Sullivan, D. (1974). Genetics of homotopy theory and the Adams conjecture. Annals of Math. 100, pp. 1-80.
[13] Fakir, S. (1970). Monade idempotente associeé à une monade. C.R. Acad. Sci. Paris 270, pp. A99-A101.
[14] Lambek, J. (1972). Localization and completion. Journal of Pure and Applied Algebra 2, pp. 343-370.
[15] Lambek, J. and Rattray, B.A. (1973). Localization at injectives in complete categories. Proc. Amer. Math. Soc. 41, pp. 1-9.
[16] Lambek, J. and Rattray, B.A. (1974). Localization and codensity triples. Comm. in Algebra 1, pp. 145-164.
[17] Lambek, J. and Rattray, B.A. (1975). Localization and duality in additive categories. Houston J. of Math. 1, pp. 87-100.

FINITARY AUTOMORPHISMS AND INTEGRAL HOMOLOGY

J. Roitberg
Hunter College, CUNY, New York, N. Y., U.S.A.

1. The notion of finitary automorphism (or pseudo-identity) of a group was introduced by J. M. Cohen in [C]. We recall the definition.

Definition. *An automorphism* $\phi: G \to G$ *of a group* G *is finitary if each element* $x \in G$ *lies in a finitely generated subgroup* $K = K_x$ *of* G *such that* ϕ *restricts to an automorphism* $\phi|K: K \cong K$.

Obviously, any automorphism of a finitely generated group G is automatically finitary. The following two examples of nonfinitary automorphisms should serve to illuminate the concept.

Example 1. Let Γ be an arbitrary nontrivial group and let $G = \bigoplus_{n=-\infty}^{\infty} \Gamma_n$, Γ_n denoting a 'copy' of Γ, $n \in Z$. If $\phi: G \to G$ is the 'shift' automorphism, carrying each Γ_n isomorphically onto Γ_{n+1} , then ϕ is a nonfinitary automorphism of G.

Example 2. Let G be the additive group of dyadic rationals, consisting of those fractions having denominator a power of 2. If $\phi: G \to G$ is 'multiplication by 2' then ϕ is a nonfinitary automorphism of G.

In [C], Cohen, whose sole concern was with abelian groups G, established a spectral sequence result with applications to fibrations of simply-connected spaces. Motivated partially by the desire to extend Cohen's result from the simply-connected to the quasi-nilpotent case ([HR1]), the authors in [HR2] develop the theory of finitary automorphisms of locally nilpotent groups G and in [CHR] initiate a theory of finitary automorphisms in the homotopy category. The following two results set the stage for the problem to be studied in the present paper; the first is Theorem 6 of [HR2], the second is a special case of one of the main theorems in [CHR] concerning nilpotent spaces.

Theorem 1. *If* $\phi: G \to G$ *is an endomorphism (not assumed to be an automorphism) of a nilpotent group, then* ϕ *is a finitary automorphism* $\Leftrightarrow$ *the induced endomorphism* $\phi_{ab}: G_{ab} \to G_{ab}$ *on the abelianization of* G *is a finitary automorphism.*

Theorem 2. *If* $\phi: G \to G$ *is an endomorphism of a nilpotent group, then* ϕ *is a finitary automorphism* $\Leftrightarrow$ *the induced endomorphism* $\phi_*: H_*G \to H_*G$ *on the integral homology of* G *is a finitary automorphism.*

With regard to Theorem 1, it is easily seen that the implication $\Rightarrow$ is valid for an arbitrary group G. The converse implication $\Leftarrow$ fails for general G since ϕ_{ab} being a finitary automorphism does not guarantee that ϕ is an automorphism. Taking Γ to be a nontrivial perfect group (that is, $\Gamma_{ab} = 1$) and $\phi: G \to G$ as in Example 1, we see that even if ϕ is an automorphism, it may not be finitary merely because ϕ_{ab} is finitary.

In a different vein, it may be noted that even if G is nilpotent, ϕ need not be an automorphism merely because ϕ_{ab} is an automorphism. In fact, let N be a nonabelian nilpotent group,

$$\Gamma_n = \begin{cases} N & \text{if } n \le 0 \\ N_{ab} & \text{if } n > 0 \end{cases},$$

$G = \bigoplus_{n=-\infty}^{\infty} \Gamma_n$ and $\phi: G \to G$ the obvious shift map. Plainly G is nilpotent, ϕ is not an automorphism but ϕ_{ab} is an automorphism. Of course, ϕ_{ab} is the type of automorphism met in Example 1 and is not finitary.

As for Theorem 2, again the implication $\Leftarrow$ fails for general G; indeed, we may reason as above for Theorem 1, replacing 'perfect' by 'acyclic' (cf. [BDH]). In contrast to the situation noted above with respect to Theorem 1, however, we remark that for nilpotent G, ϕ must be an automorphism if ϕ_* is an automorphism.

For the remainder of the paper we focus attention on the question of the validity of the implication $\Rightarrow$ in Theorem 2 for general G. We have:

Theorem. *There exists a finitely presented group* G *and an automorphism* $\phi: G \to G$ *(necessarily finitary) such that* $\phi_*: H_*G \to H_*G$ *is not a finitary automorphism.*

The proof of this theorem together with some additional commentary will occupy the next section.

It is a great pleasure for me to dedicate this paper to Peter Hilton on the occasion of his sixtieth birthday. The observations contained herein are, in part, an outgrowth of work resulting from our collaboration, a collaboration which I trust will continue for many years to come.

2. We show that the group G constructed by Stallings in [S] supports an automorphism of the desired kind.

Following the notation in [S], we begin with a free group on two generators $A = \langle a,b \rangle$. Let $\phi_A: A \to A$ denote conjugation by the element a. If B is the normal subgroup of A generated by b, then we have the presentation $B = \langle a^n b a^{-n},\ n \in Z \rangle$, that is, B is freely generated by the indicated elements. We note that ϕ_A restricts to an automorphism ϕ_B of B. Hence, if C is the generalized free square $A \underset{B}{*} A$, ϕ_A naturally induces a unique automorphism $\phi_C: C \to C$; writing

$$C = \langle a_1, a_2, b;\ a_1^n b a_1^{-n} = a_2^n b a_2^{-n},\ n \in Z \rangle ,$$

ϕ_C is determined by the assignment $a_1 \to a_1$, $a_2 \to a_2$, $b \to a_1 b a_1^{-1}$ $(= a_2 b a_2^{-1})$. The next step is to embed C in the finitely presented group

$$D = \langle a_1, a_2, b, x;\ [x,a_1] = 1 = [x,a_2],\ xbx^{-1} = a_1 b a_1^{-1} = a_2 b a_2^{-1} \rangle$$

and to observe that if $\phi_D: D \to D$ denotes conjugation by the element x, then $\phi_D | C = \phi_C$. Finally, we take G to be the generalized free square $D \underset{C}{*} D$ and let $\phi = \phi_G: G \to G$ be the unique automorphism induced by ϕ_D.

We claim that $\phi_*: H_3 G \to H_3 G$ is a nonfinitary automorphism. To this end, we begin by observing that $(\phi_B)_*: H_1 B \to H_1 B$ is a nonfinitary automorphism. In fact, we may identify $H_1 B$ with B_{ab}, $(\phi_B)_*: H_1 B \to H_1 B$ with $(\phi_B)_{ab}: B_{ab} \to B_{ab}$ and note that $(\phi_B)_{ab}$ is precisely the automorphism discussed in Example 1 with $\Gamma = Z$. Now $(\phi_A)_*: H_1 A \to H_1 A$ is certainly a finitary automorphism (actually, $(\phi_A)_* =$ identity since ϕ_A is an inner automorphism). By applying [HR2; Corollary 3] to the map of Mayer-Vietoris sequences

$$\begin{array}{ccccc} H_2 C & \longrightarrow & H_1 B & \longrightarrow & H_1 A \oplus H_1 A \\ \downarrow (\phi_C)_* & & \downarrow (\phi_B)_* & & \downarrow (\phi_A)_* \oplus (\phi_A)_* \\ H_2 C & \longrightarrow & H_1 B & \longrightarrow & H_1 A \oplus H_1 A \end{array} ,$$

we conclude that $(\phi_C)_*: H_2 C \to H_2 C$ is a nonfinitary automorphism. Finally, $(\phi_D)_*: H_2 D \to H_2 D$ is a finitary automorphism (again, $(\phi_D)_*$ = identity), so by once again applying [HR2; Corollary 3] to the map of Mayer-Vietoris sequences

$$\begin{array}{ccccc} H_3G & \longrightarrow & H_2C & \longrightarrow & H_2D \oplus H_2D \\ \downarrow \phi_* & & \downarrow (\phi_C)_* & & \downarrow (\phi_D)_* \oplus (\phi_D)_* \\ H_3G & \longrightarrow & H_2C & \longrightarrow & H_2D \oplus H_2D \end{array} ,$$

we conclude that $\phi_*: H_3G \to H_3G$ is a nonfinitary automorphism, thereby completing the proof of the theorem.

We have seen that the automorphism $\phi_*: H_3G \to H_3G$ is suitably manufactured from the type of nonfinitary automorphism presented in Example 1. As was kindly pointed out to me by Gilbert Baumslag, it is also possible to begin from the type of nonfinitary automorphism presented in Example 2 and fashion a suitable example. We very briefly outline the construction.

Take $A = \langle a,b;\ aba^{-1} = b^2 \rangle$, with ϕ_A conjugation by the element a. Let B be the normal subgroup of A generated by b. It is readily checked that B is isomorphic to the additive group of dyadic rationals and that ϕ_A restricts to an automorphism ϕ_B of B which may be identified with 'multiplication by 2'. The constructions of C, D, G and ϕ_C, ϕ_D, $\phi = \phi_G$ are then performed in precisely the same manner as in Stallings' example.

Since it is known ([BDH]) that there is a functorial embedding of any finitely generated group G into a finitely generated acyclic group $\mathcal{A}G$, it follows that any finitely generated group G admits a finitely generated 'suspension' $\Sigma G = \mathcal{A}G \underset{G}{*} \mathcal{A}G$. As this Σ construction is likewise functorial, any automorphism $\phi: G \to G$ induces a unique automorphism $\Sigma\phi: \Sigma G \to \Sigma G$ and a simple Mayer-Vietoris argument shows that ϕ_* is finitary $\Leftrightarrow$ $(\Sigma\phi)_*$ is finitary. By applying the iterated suspension operation to the $\phi: G \to G$ of the theorem (or even to $\phi_C: C \to C$), we thus easily deduce the following corollary to the theorem.

<u>Corollary</u>. *There exists a finitely generated group* K *of arbitrarily large homological connectivity and an automorphism* $\psi: K \to K$ *such that* $\psi_*: H_*K \to H_*K$ *is not a finitary automorphism.*

If it were known that there is a functorial embedding of any finitely presented group into a finitely presented acyclic group, then we could, of course, strengthen the corollary suitably. There is a process (cf. [BDM]) for embedding a finitely presented group into a finitely presented acyclic group but this process is definitely not a functorial one.

To close, we wish to raise a further problem. The various groups considered in this section appear to be rather complicated in many ways. It would seem fair to say, for instance, that they are very far from being nilpotent. Thus, it might be interesting to find an example validating the theorem with the group G, say, a solvable group. In [BD], the authors construct a pair of finitely presented metabelian groups, each with infinitely generated integral homology groups in most degrees, so these groups are perhaps good candidates. Unfortunately, it is not at all clear how to concoct a suitable automorphism on either of these groups.

Bibliography

[BD] Baumslag, G. & Dyer, E. (1982). The integral homology of finitely generated metabelian groups, I. Amer. J. Math. 104, 173-182.

[BDH] Baumslag, G., Dyer, E. & Heller, A. (1980). The topology of discrete groups. J. Pure & Appl. Algebra 16, 1-47.

[BDM] Baumslag, G., Dyer, E. & Miller, C. F. On the integral homology of finitely presented groups. To appear in Topology; see also (1981) Bull. Amer. Math. Soc. 4, 321-324.

[CHR] Castellet, M., Hilton, P. & Roitberg, J. On pseudo-identities II. In preparation.

[C] Cohen, J. M. (1968). A spectral sequence automorphism theorem; applications to fibre spaces and stable homotopy. Topology 7, 173-177.

[HR1] Hilton, P. & Roitberg, J. (1976). On the Zeeman comparison theorem for the homology of quasi-nilpotent fibrations. Quart. J. Math. 27, 433-444.

[HR2] Hilton, P. & Roitberg, J. On pseudo-identities I. To appear.

[S] Stallings, J. (1963). A finitely presented group whose 3-dimensional integral homology is not finitely generated. Amer. J. Math. 85, 541-543.

FINITE GROUP ACTIONS ON GRASSMANN MANIFOLDS

Henry H. Glover and Guido Mislin

Ohio State University and ETH Zürich

(Dedicated to Peter Hilton on the occasion of his sixtieth birthday)

INTRODUCTION.

Let $\rho : G \to GL_n(\mathbb{C})$ denote a complex representation of the finite group G. If $\sigma \in Gal(\mathbb{C}/\mathbb{Q})$ is a field automorphism of $\mathbb{C}$ then by applying σ to the entries of a matrix one obtains an induced automorphism $\sigma_* : GL_n(\mathbb{C}) \to GL_n(\mathbb{C})$. We denote the composite map $\sigma_* \circ \rho$ by ρ^σ and we call it a Galois conjugate of ρ. Since G is finite, two representations ρ_1 and ρ_2 are equivalent $(\rho_1 \sim \rho_2)$ if and only if their characters χ_{ρ_1} and χ_{ρ_2} agree. In particular $\rho^\sigma \sim \rho^\tau$ if σ and τ act in the same way on $|G|$-th roots of unity in $\mathbb{C}$.

Let $\mu \subset \mathbb{C}$ be the group of roots of unity and write $\deg : Gal(\mathbb{C}/\mathbb{Q}) \to Aut(\mu)$ for the map given by restricting an automorphism of $\mathbb{C}$ to μ. We use the notation $\deg(\sigma) \equiv k \bmod m$, m a natural number, if σ acts on m-th roots of unity by the k-power map. Note that $\deg : Gal(\mathbb{C}/\mathbb{Q}) \to Aut(\mu)$ is surjective and $Aut(\mu) \cong \varprojlim Aut(\mu_m)$, $\mu_m \subset \mu$ the group of m-th roots of unity. As usual, we denote by $\rho_1 \otimes \rho_2$ the (equivalence class) of the tensor product representation of ρ_1 and ρ_2.

If $\rho : G \to GL_{s+t}(\mathbb{C})$ is a representation, then G acts on the Grassmann manifold $\mathbb{C}_{s,t}$ of s-dimensional linear subspaces of $\mathbb{C}^{s+t}$; we write $\mathbb{C}_{s,t}(\rho)$ for the corresponding G-manifold. Clearly, if ρ_1 and ρ_2 are projectively equivalent (i.e., if there exists a one dimensional representation $\lambda : G \to \mathbb{C}^*$ such that $\rho_1 \sim \lambda \otimes \rho_2$), there is an equivariant homeomorphism $\mathbb{C}_{s,t}(\rho_1) \cong \mathbb{C}_{s,t}(\rho_2)$. If $\bar{\rho}$ denotes the complex-conjugate of the representation ρ, then complex conjugation induces an equivariant homeomorphism $\mathbb{C}_{s,t}(\rho) \cong \mathbb{C}_{s,t}(\bar{\rho})$. Liulevicius proved [16] that conversely if there is an equivariant map $\mathbb{C}_{s,t}(\rho_1) \to \mathbb{C}_{s,t}(\rho_2)$, which is a homotopy equivalence on the underlying space, then there exists a one dimensional representation λ such that either $\rho_1 \sim \lambda \otimes \rho_2$ or $\bar{\rho}_1 \sim \lambda \otimes \rho_2$.

We will study more generally the connection between equivariant maps $f : \mathbb{C}_{s,t}(\rho_1) \to \mathbb{C}_{s,t}(\rho_2)$ and the representations ρ_1 and ρ_2; Liulevicius' result will correspond to the case where f is a homotopy equivalence. We will also deal with the case $s = t$, for which one needs Hoffman's result [12] on the automorphisms of the cohomology algebra of $\mathbb{C}_{s,s}$ (cf. Lemma 1.5). Our results should be compared with similar results on finite group action on a sphere (see Atiyah-Tall [2] and Lee-Wasserman [14]). The proofs are very different however; we follow essentially the method initiated by Liulevicius in [16].

It is convenient to use the following notion of degree for a self map $f : \mathbb{C}_{s,t} \to \mathbb{C}_{s,t}$. Since $H^2(\mathbb{C}_{s,t}; \mathbb{Z}) \cong \mathbb{Z}$, there is a unique integer $\deg(f)$ corresponding to the induced homomorphism of this second homology group. The Kähler class, which generates $H^2(\mathbb{C}_{s,t}; \mathbb{Z})$, has maximal cup length; thus a map of degree ± 1 is necessarily a homotopy equivalence (cf. Brewster-Homer [5]).

Our main result can be stated as follows.

Main Theorem: Let s and t be natural numbers and let $\rho_1, \rho_2 : G \to GL_{s+t}(\mathbb{C})$ be representations of the finite group G. Suppose there exists an equivariant map $f : \mathbb{C}_{s,t}(\rho_1) \to \mathbb{C}_{s,t}(\rho_2)$ of degree prime to $|G|$. Then

A) There exists a one dimensional representation λ of G such that ρ_1 and $\lambda \otimes \rho_2$ are Galois conjugate. If $\sigma \in \mathrm{Aut}(\mathbb{C}/\mathbb{Q})$ denotes an automorphism such that $\rho_1 \sim \lambda \otimes \rho_2$, then σ and f are related as follows.

B1) If $s \neq t$, or if $s = t$ and f induces a grading map in cohomology (i.e. $f^*x = \deg(f)^n x$ for $x \in H^{2n}$), then $\deg(\sigma) \equiv \deg(f) \bmod |G|$.

B2) If $s = t$ and f does not induce a grading map in cohomology, then $\deg(\sigma) \equiv -\deg(f) \bmod |G|$.

The maps f considered in our Main Theorem are in general non-linear in the sense that they are not induced from linear maps of $\mathbb{C}^{s+t}$; a linear f has necessarily degree 0 or 1.

If the representation ρ is equivalent to a representation defined over $\mathbb{Q}$, then $\rho \sim \rho^{\sigma}$ for all $\sigma \in \mathrm{Gal}(\mathbb{C}/\mathbb{Q})$. This is for instance the case for all representations of symmetric groups. The following is then an immediate consequence of our Main Theorem.

<u>Corollary</u>: Let $f : \mathbb{C}_{s,t}(\rho_1) \to \mathbb{C}_{s,t}(\rho_2)$ be an equivariant map between Grassmann G-manifolds where G is a symmetric group. If the degree of f is prime to $|G|$ then there is a linear G-homeomorphism $\mathbb{C}_{s,t}(\rho_1) \to \mathbb{C}_{s,t}(\rho_2)$.

Our paper is organized as follows. In section one we prove the Main Theorem. We show in section two how to construct equivariant maps in the case of actions on $\mathbb{C}_{1,n+1}$, that is, n-dimensional complex projective space $\mathbb{C}P^n$. It is somewhat mysterious how to construct such equivariant maps in the general situation. However, it is possible to formulate our Main Theorem in a more general way (section three). The theorem has then a converse, stating that if ρ_1 is Galois conjugate to ρ_2, then there is a map $f : \mathbb{C}_{s,t}(\rho_1) \to \mathbb{C}_{s,t}(\rho_2)$ of degree prime to $|G|$, which is homotopic to an equivariant map at "each prime which divides $|G|$" (cf. Theorem 3.3).

1. LIULEVICIUS' USE OF REPRESENTATION RINGS.

We will write $R(K)$ for the complex representation ring of the compact Lie group K. The Adams operations $\psi^k : R(K) \to R(K)$ are then defined in the usual way. If G is a finite group, then $\sigma \in \mathrm{Gal}(\mathbb{C}/\mathbb{Q})$ gives rise to an automorphism $R(\sigma) : R(G) \to R(G)$, induced from mapping a representation ρ to ρ^σ. The following lemma states the well known relationship between ψ^k and $R(\sigma)$ (a proof may be found in Eckmann-Mislin [7]).

Lemma 1.1. If G denotes a finite group and k is an integer prime to $|G|$, then for every $\sigma \in \mathrm{Gal}(\mathbb{C}/\mathbb{Q})$ such that $\deg(\sigma) \equiv k \bmod |G|$ one has

$$\psi^k = R(\sigma) : R(G) \to R(G).$$

Let $\rho : G \to U(s + t)$ be a unitary representation of the finite group G. Then $\rho^* : R(U(s + t)) \to R(G)$ makes $R(G)$ into an $R(U(s + t))$ - algebra. Similarly, the usual inclusion $U(s) \times U(t) \to U(s+t)$ turns $R(U(s) \times U(t))$ into an $R(U(s + t))$ - algebra. Thus we can form the tensor product

$$RG \otimes_{R(U(s + t))} R(U(s) \times U(t))$$

which we will denote by $(RG \otimes R(U(s) \times U(t)))_\rho$. This tensor product has an obvious $R(G)$ - algebra structure.

To solve our problem on Grassmann G-manifolds, we will henceforth assume that all representations ρ of G, which we consider, are unitary; there is of course no loss in generality in doing so. If $\rho : G \to U(s+t)$ is such a representation, then the usual identification $\mathbb{C}_{s,t} \cong U(s+t)/U(s) \times U(t)$ allows us to compute the equivariant K-theory of $\mathbb{C}_{s,t}(\rho)$ as follows (compare Snaith [18] and Liulevicius [16]).

<u>Lemma 1.2</u>. If the finite group G acts on $\mathbb{C}_{s,t}$ via $\rho : G \to U(s+t)$, then there is a natural $R(G)$ - algebra isomorphism

$$A(\rho) : (R(G) \otimes R(U(s) \times U(t)))_{\rho} \to K^{\circ}_{G}(\mathbb{C}_{s,t}(\rho)).$$

If $\phi : (RG \otimes R(U(s) \times U(t)))_{\rho_1} \to (RG \otimes R(U(s) \times U(t)))_{\rho_2}$ is a map of RG - algebras then we define the degree of ϕ as follows. Using the augmentation $\dim : RG \to \mathbb{Z}$ and tensoring with $\mathbb{Q}$ gives an induced map

$$\overline{\phi} : \mathbb{Q} \otimes_{R(U(s+t))} R(U(s) \times U(t)) \to \mathbb{Q} \otimes_{R(U(s+t))} R(U(s) \times U(t))$$

Since $\mathbb{Q} \otimes_{R(u(s+t))} R(U(s) \times U(t))$ is isomorphic to $K^{\circ}(\mathbb{C}_{s,t}) \otimes \mathbb{Q}$ we obtain from $\overline{\phi}$, using the Chern character, an endomorphism ϕ^* of $H^*(\mathbb{C}_{s,t}; \mathbb{Q})$. This endomorphism might fail to preserve the gradation; we define the degree of ϕ in general to be the degree of the composite of $\phi^* : H^2(\mathbb{C}_{s,t}; \mathbb{Q}) \to H^*(\mathbb{C}_{s,t}; \mathbb{Q})$ with the projection $H^*(\mathbb{C}_{s,t}; \mathbb{Q}) \to H^2(\mathbb{C}_{s,t}; \mathbb{Q})$.

The following three lemmas are crucial for the proof of our theorem. They will permit us to reduce the degree k situation to the degree one situation as considered in [16].

Lemma 1.3. Let $\rho : G \to U(s + t)$ be a representation of the finite group G and let $\sigma \in \mathrm{Gal}(\mathbb{C}/\mathbb{Q})$ be a Galois automorphism. If $k \in \mathbb{Z}$ is such that $\deg(\sigma) \equiv k \bmod |G|$ then there is a map of RG-algebras

$$\theta_k : (RG \otimes R(U(s) \times U(t)))_{\rho^\sigma} \to (RG \otimes R(U(s) \times U(t)))_\rho$$

of degree k, given by $\alpha \otimes \beta \to \alpha \otimes \psi^k \beta$. Moreover, θ_k induces a grading map θ_k^* in $H^*(\mathbb{C}_{s,t}; \mathbb{Q})$.

Proof. The map θ_k is induced from the following diagram, whose rows have as push-outs the algebras in question:

$$\begin{array}{ccccc}
R(G) & \xleftarrow{(\rho^\sigma)^*} & R(U(s+t)) & \longrightarrow & R(U(s) \times U(t)) \\
\downarrow \mathrm{Id} & & \downarrow \psi^k & & \downarrow \psi^k \\
R(G) & \xleftarrow{\rho^*} & R(U(s+t)) & \longrightarrow & R(U(s) \times U(t)).
\end{array}$$

Note that k is necessarily prime to $|G|$ since the degree of σ is a unit mod $|G|$. The commutativity of the diagram follows from Lemma 1.1 and the naturality of the ψ-operations. The induced map

$$\overline{\theta}_k : \mathbb{Q} \otimes_{R(U(s+t))} R(U(s) \times U(t)) \to \mathbb{Q} \otimes_{R(U(s+t))} R(U(s) \times U(t))$$

corresponds under the natural isomorphism

$$\mathbb{Q} \otimes_{R(U(s+t))} R(U(s) \times U(t)) \cong K^\circ(\mathbb{C}_{s,t}) \otimes \mathbb{Q}$$

to the Adams operation ψ^k on vector bundles, which correspond via the Chern character to a grading map of degree k on $H^*(\mathbb{C}_{s,t}; \mathbb{Q})$, as is well known. Thus θ_k has degree k and induces a grading map θ_k^* in $H^*(\mathbb{C}_{s,t}; \mathbb{Q})$.

<u>Lemma 1.4</u>. The map θ_k of Lemma 1.3 induces an isomorphism

$$\mathbb{Q} \otimes \theta_k : \mathbb{Q} \otimes (RG \otimes R(U(s) \times U(t)))_{\rho^\sigma} \to \mathbb{Q} \otimes (RG \otimes R(U(s) \times U(t)))_\rho .$$

<u>Proof</u>. It suffices to show that $\mathbb{Q} \otimes \theta_k$ is injective, since its domain and range are finite dimensional $\mathbb{Q}$-vector spaces of the same dimension (the dimension is $N \cdot \dim_{\mathbb{Q}}(\mathbb{Q} \otimes RG)$ where $N = (s + t)!/(s!t!)$). Choose $\tau \in \mathrm{Gal}(\mathbb{C}/\mathbb{Q})$ such that $\deg(\sigma\tau) \equiv 1 \bmod |G|$ and thus $(\rho^\sigma)^\tau \sim \rho$. If $m \in \mathbb{Z}$ is such that $\deg(\tau) \equiv m \bmod |G|$ then $\mathbb{Q} \otimes \theta_{mk} = (\mathbb{Q} \otimes \theta_m)(\mathbb{Q} \otimes \theta_k)$ is an endomorphism of $\mathbb{Q} \otimes (RG \otimes R(U(s) \times U(t)))_\rho \cong K^\circ_G(\mathbb{C}_{s,t}(\rho)) \otimes \mathbb{Q}$. By Slominska [17], $K^\circ_G(\mathbb{C}_{s,t}(\rho)) \otimes \mathbb{Q}$ injects into $\prod_{H \subseteq G}(K^\circ(\mathbb{C}_{s,t}(\rho)^H) \otimes RH \otimes \mathbb{Q})$ and for every component of the map into the product one has a commutative diagram

$$\begin{array}{ccc}
\mathbb{Q} \otimes (RG \otimes R(U(s) \times U(t)))_\rho & \longrightarrow & \mathbb{Q} \otimes RH \otimes K^\circ(\mathbb{C}_{s,t}(\rho)^H) \\
\Big\downarrow \mathbb{Q} \otimes \theta_{mk} & & \Big\downarrow 1 \otimes 1 \otimes \psi^{mk} \\
\mathbb{Q} \otimes (RG \otimes R(U(s) \times U(t)))_\rho & \longrightarrow & \mathbb{Q} \otimes RH \otimes K^\circ(\mathbb{C}_{s,t}(\rho)^H)
\end{array}$$

But $1 \otimes \psi^{mk} : \mathbb{Q} \otimes K^\circ(\mathbb{C}_{s,t}(\rho)^H) \to \mathbb{Q} \otimes K^\circ(\mathbb{C}_{s,t}(\rho)^H)$ is obviously an isomorphism, as one can see by passing to ordinary cohomology with rational coefficients, using the Chern character. Thus, $\mathbb{Q} \otimes \theta_{mk}$ is monic and so is $\mathbb{Q} \otimes \theta_k$.

Next, we will describe the structure of endomorphisms and automorphisms of the cohomology algebra of $\mathbb{C}_{s,t}$ which we will need. The corresponding results are scattered around in the literature. Early results are due to Glover-Homer [10]. Brewster [4] classified all endomorphisms of non-zero degree of $H^*(\mathbb{C}_{s,t}; \mathbb{Z})$ if $s \neq t$; his method applies equally well to the cohomology with coefficients in a field of characteristic 0. The automorphisms of $H^*(\mathbb{C}_{s,s}; \mathbb{Z})$ were determined by

Hoffman in [12]; again, his methods yield a corresponding result in case of coefficients in a field of characteristic 0. A discussion of the automorphism conjecture for generalized flag manifolds and its connection to rational homotopy theory is presented in Glover-Mislin [11]. For our Main Theorem we need the endomorphism result for rational coefficients; this is discussed entirely in Brewster-Homer [5]. We will also need the corresponding result for coefficients in the field of p-adic numbers (cf. section three). The sources mentioned above imply the following general result.

<u>Lemma 1.5</u>. Let F be a field of characteristic 0 and let $\phi : H^*(\mathbb{C}_{s,t}; F) \to H^*(\mathbb{C}_{s,t}; F)$ be an F-algebra endomorphism of degree $k \in F$, $k \neq 0$. Then ϕ is an automorphism and the following holds.

(i) If $s \neq t$, ϕ is a grading map (i.e., $\phi x = k^n x$ for $x \in H^{2n}$).

(ii) If $s = t$, then either ϕ or $\phi \circ \kappa^*$ is a grading map, where $\kappa^* : H^*(\mathbb{C}_{s,s}; F) \to H^*(\mathbb{C}_{s,s}; F)$ denotes the involution of degree -1 induced from mapping an s-plane in $\mathbb{C}^{2s}$ to the orthogonal complement with respect to some fixed unitary metric on $\mathbb{C}^{2s}$.

<u>Proposition 1.6</u>. Let $f : \mathbb{C}_{s,t}(\rho_1) \to \mathbb{C}_{s,t}(\rho_2)$ be an equivariant map of degree k prime to $|G|$. Then there exists an equivariant map $\tilde{f} : \mathbb{C}_{s,t}(\rho_1) \to \mathbb{C}_{s,t}(\rho_2)$ and $\sigma \in \mathrm{Gal}(\mathbb{C}/\mathbb{Q})$ such that $\deg \sigma \equiv \varepsilon k \bmod |G|$, where $\varepsilon = +1$ if f induces a grading map in cohomology, and $\varepsilon = -1$ in the other case, such that the following diagram is commutative

$$\begin{array}{ccc}
\mathbb{Q} \otimes (RG \otimes R(U(s) \times U(t)))_{\rho_2} & \xrightarrow{(\mathbb{Q} \otimes \theta_{\varepsilon k})^{-1} \circ \tilde{f}^!} & \mathbb{Q} \otimes (RG \otimes R(U(s) \times U(t)))^{\sigma}_{\rho_1} \\
{\scriptstyle 1 \otimes \dim \otimes 1}\downarrow & & \downarrow{\scriptstyle 1 \otimes \dim \otimes 1} \\
\mathbb{Q} \otimes_{R(U(s+t))} R(U(s) \times U(t)) & \xrightarrow{\text{id}} & \mathbb{Q} \otimes_{R(U(s+t))} R(U(s) \times U(t))
\end{array}$$

Proof. If f induces a grading map, we take $\tilde{f} = f$ and apply Lemma 1.3 and 1.4 respectively. If f does not induce a grading map, we are in case (ii) of Lemma 1.5 and thus $f \circ \kappa$ is a grading map (of degree $-k$). We can choose $\kappa : \mathbb{C}_{s,s}(\rho_1) \to \mathbb{C}_{s,s}(\rho_1)$ to be equivariant, by mapping an s-plane to the orthogonal complement with respect to a ρ_1-invariant metric on $\mathbb{C}^{2s}$. The assertion of our Proposition then follows in this case by taking $\tilde{f} = f \circ \kappa$ and noting that $\deg(\kappa) = -1$.

The proof of our Main Theorem can now be completed by applying Liulevicius' result [16]. From there it follows that the existence of a diagram as considered in Proposition 1.6 implies that there is a one dimensional representation $\lambda : G \to \mathbb{C}^*$ such that $\rho_1^{\sigma} \sim \lambda \otimes \rho_2$. Since $\deg(\sigma) \equiv \deg(f) \bmod |G|$ if f induces a grading map, B1 of the Main Theorem follows by Lemma 1.5. Similarly, B2 follows by noting that $\deg(\sigma) \equiv -\deg(f) \bmod |G|$ if f does not induce a grading map, which can only occur in case $s = t$.

2. EQUIVARIANT MAPS ON PROJECTIVE SPACES.

We will discuss the construction of equivariant maps as considered in the Main Theorem for the case of $\mathbb{C}_{1,n+1} = \mathbb{C}P^n$, complex projective n-space.

Lemma 2.1. Let $\rho : G \to GL_{n+1}(\mathbb{C})$ be a representation of a finite group G. Assume that ρ is a sum of representations which are induced from one dimensional representations of subgroups of G. If $\sigma \in \mathrm{Gal}(\mathbb{C}/\mathbb{Q})$ is a Galois automorphism and if $\deg(\sigma) \equiv k \bmod |G|$, $k \in \mathbb{Z}$, then there exists an equivariant map

$$f : \mathbb{C}P^n(\rho) \to \mathbb{C}P^n(\rho^\sigma)$$

of degree k.

Proof. Since ρ is a sum of representations induced from one dimensional representations, there exists a basis $\{e_1, \ldots, e_{n+1}\}$ of $\mathbb{C}^{n+1}$ such that with respect to this basis

$$\rho(g)(z_1, \ldots, z_{n+1}) = (h_1 z_{\pi(1)}, \ldots, h_{n+1} z_{\pi(n+1)})$$

where $h_1, \ldots, h_{n+1} \in \mathbb{C}$ and π a permutation (cf. Dornhoff [6]); π and $h_1, \ldots, h_{n+1}$ all depend on $g \in G$, and $h_1, \ldots, h_{n+1}$ are necessarily $|G|$-th roots of unity. Since σ acts by the k-power map on $|G|$-th roots of unity, we infer

$$\rho^\sigma(g)(z_1, \ldots, z_{n+1}) = (h_1^k z_{\pi(1)}, \ldots, h_{n+1}^k z_{\pi(n+1)}).$$

Let P^k denote the map $\mathbb{C}^{n+1} \to \mathbb{C}^{n+1}$ given by $\Sigma z_j e_j \longmapsto \Sigma z_j^k e_j$. Then, by construction $\rho^\sigma(g) \circ P^k = P^k \circ \rho(g)$ for all $g \in G$. Therefore, the induced map $\bar{P}^k = f : \mathbb{C}P^n(\rho) \to \mathbb{C}P^n(\rho^\sigma)$ is an equivariant map, and the degree of f is obviously equal to k.

Using this Lemma, we obtain the following partial converse to our Main Theorem.

Theorem 2.2. Let $\rho_1, \rho_2 : G \to GL_{n+1}(\mathbb{C})$ be two representations of a finite nilpotent group G. Suppose there exists $\sigma \in \mathrm{Gal}(\mathbb{C}/\mathbb{Q})$ and $\lambda : G \to \mathbb{C}^*$ such that $\rho_1^\sigma \sim \lambda \otimes \rho_2$. Then, for every $k \in \mathbb{Z}$ with $k \equiv \deg(\sigma) \bmod |G|$, there exists an equivariant map

$$f : \mathbb{C}P^n(\rho_1) \to \mathbb{C}P^n(\rho_2)$$

with $\deg(f) = k$.

Proof. Since G is nilpotent, ρ_1 is equivalent to a sum of representations induced from one dimensional representations of subgroups [6]. By Lemma 2.1 we can therefore construct an equivariant map $g : \mathbb{C}P^n(\rho_1) \to \mathbb{C}P^n(\rho_1^\sigma)$ of degree k. Since $\rho_1^\sigma \sim \lambda \otimes \rho_2$, there is an equivariant homeomorphism $h : \mathbb{C}P^n(\rho_1^\sigma) \to \mathbb{C}P^n(\rho_2)$ of degree one. Thus $f = h \circ g$ is the desired map of degree k.

3. EQUIVARIANT MAPS AT A PRIME p.

We consider G - CW complexes X in the sense of Illman [13]. If G is a finite group, then the localization and completion of Bousfield-Kan [3] may be used to define functorially localized G - CW complexes X_p and p-completed G - CW complexes $\hat{X}_p$, p a prime. If the fixed point sets X^H have all components nilpotent and of finite type, then the canonical map $X \to \hat{X}_p$ will induce isomorphims $H_*(X^H; \mathbb{Z}/p) \to H_*((\hat{X}_p)^H; \mathbb{Z}/p)$ for all $H \subset G$, thus $(X^H)\hat{}_p \simeq (\hat{X}_p)^H$.

Definition 3.1. A map $f : X \to Y$ between G - CW complexes is called equivariant at p, if $\hat{f}_p : \hat{X}_p \to \hat{Y}_p$ is homotopic to an equivariant map.

Such a map $f : X \to Y$ which is equivariant at p will give rise to an induced map

$$\hat{f}_p^! : K_G^\circ(\hat{Y}_p) \to K_G^\circ(\hat{X}_p).$$

The notation $\hat{f}_p^!$ is however somewhat misleading since the induced map may depend on the choice of the equivariant representative of $\hat{f}_p$; we will assume in the sequel that such a choice has been made once and for all. Note that $K^*(\hat{X}_p; \hat{\mathbb{Z}}_p) \cong \varprojlim K^*(\hat{X}_p; \mathbb{Z}/p^n\mathbb{Z})$ since the groups $K^*(\hat{X}_p; \mathbb{Z}/p^n\mathbb{Z})$ have a natural compact topology; similarly for K_G^*. Also, if X is finite and nilpotent, then $K^*(\hat{X}_p; \hat{\mathbb{Z}}_p) \cong K^*(X) \otimes \hat{\mathbb{Z}}_p$. Thus the ordinary Chern character gives rise in this case to an isomorphism

$$K^{\circ}(\hat{X}_p; \hat{\mathbb{Z}}_p) \otimes \mathbb{Q} \cong H^{ev}(\hat{X}_p; \hat{\mathbb{Z}}_p) \otimes \mathbb{Q}.$$

Similarly, the equivariant Chern character of Slominska [17] can easily be adapted to the p-completed situation. Let us write $R_G \otimes \mathbb{Q}_p$ for the coefficient system (in the sense of Bredon) given by $(R_G \otimes \mathbb{Q}_p)(G/H) = R(H) \otimes \mathbb{Q}_p$, $\mathbb{Q}_p = \hat{\mathbb{Z}}_p \otimes \mathbb{Q}$ the field of p-adic numbers. Then for X a G - CW complex with G finite and each connected component of X^H a finite nilpotent complex, there is a natural isomorphism

$$ch_G : K_G^{\circ}(\hat{X}_p; \hat{\mathbb{Z}}_p) \otimes \mathbb{Q} \to H_G^{ev}(\hat{X}_p; R_G \otimes \mathbb{Q}_p)$$

and the canonical map

$$H_G^{ev}(\hat{X}_p; R_G \otimes \mathbb{Q}_p) \to \prod_{H \subset G} H^{ev}(\hat{X}_p^H; \mathbb{Q}_p) \underset{\mathbb{Q}_p}{\otimes} (RH \otimes \mathbb{Q}_p)$$

is injective.

The following generalization of our Main Theorem then holds.

<u>Theorem 3.2</u>. Let $\rho_1, \rho_2 : G \to GL_{s+t}(\mathbb{C})$ be representations of the finite group G. If there exists a map

$$f : \mathbb{C}_{s,t}(\rho_1) \to \mathbb{C}_{s,t}(\rho_2)$$

of degree prime to $|G|$ such that f is equivariant at all primes which divide $|G|$, then there is a one dimensional representation λ of G with the property that ρ_1 and $\lambda \otimes \rho_2$ are Galois conjugate.

Proof. Write $\mathbb{Q}_{|G|}$ for $\Pi\mathbb{Q}_p$, the product being taken over all prime divisors of G. The argument then is completely analoguous to the proof of the Main Theorem, replacing $\mathbb{Q}$ by $\mathbb{Q}_{|G|}$ at all the obvious places.

The Theorem then admits the following converse.

Theorem 3.3. Suppose $\rho_1, \rho_2 : G \to GL_{s+t}(\mathbb{C})$ are representations such that $\rho_1^\sigma \sim \lambda \otimes \rho_2$ for suitable $\sigma \in Gal(\mathbb{C}/\mathbb{Q})$ and $\lambda : G \to \mathbb{C}^*$. Then there exists a map

$$f : \mathbb{C}_{s,t}(\rho_1) \to \mathbb{C}_{s,t}(\rho_2)$$

with $\deg f \equiv \deg(\sigma) \bmod |G|$ such that f is equivariant at all primes which divide $|G|$.

Proof. The fixed point sets $\mathbb{C}^H_{s,t}$ for a linear action on $\mathbb{C}_{s,t}$ are disjoint unions of products of complex Grassmann manifolds, which are nilpotent and finite. The canonical map

$$(\mathbb{C}_{s,t}(\rho_1))\hat{}_p \to (et - \mathbb{C}_{s,t}(\rho_1))\hat{}_p$$

which one obtains by realizing the p-adic etale homotopy type of $\mathbb{C}_{s,t}(\rho_1)$ (cf. Artin-Mazur [1]) is therefore an equivariant homotopy equivalence (it induces a homology equivalence with $\mathbb{Z}/p$ coefficients on all fixed point sets).

Since $\rho_1^\sigma \sim \rho_1^\tau$ for every $\tau \in Gal(\mathbb{C}/\mathbb{Q})$ with $\deg(\tau) \equiv \deg(\sigma) \bmod |G|$, we may assume without loss of generality that there exists a large prime number N (larger than $(s + t)!$) such that $\deg(\sigma) \equiv N \bmod |G|^j$ for all j. Clearly, σ induces an equivariant map

$$\hat{\sigma}_p : (et - \mathbb{C}_{s,t}(\rho_1))\hat{}_p \to (et - \mathbb{C}_{s,t}(\rho_1^\sigma))\hat{}_p$$

and $\deg \hat{\sigma}_p \in \hat{\mathbb{Z}}_p^*$ is N for all primes p which divide $|G|$. By Friedlander [9] we can, since $(N, (s+t)!) = 1$, construct a global map

$$g : \mathbb{C}_{s,t}(\rho_1) \to \mathbb{C}_{s,t}(\rho_1^\sigma)$$

with $\hat{g}_p \simeq \hat{\sigma}_p$ for all primes p which divide $|G|$. The map g is then by definition equivariant at all primes which divide $|G|$. Since $\rho_1^\sigma \sim \lambda \otimes \rho_2$, there is an equivariant homeomorphism $h : \mathbb{C}_{s,t}(\rho_1^\sigma) \to \mathbb{C}_{s,t}(\rho_2)$ of degree 1. The map $f = h \circ g$ is then the desired map with $\deg(f) \equiv \deg(\sigma) \bmod |G|$.

Remark. Instead of considering actions on Grassmann manifolds, one could in a similar way study actions on generalized flag manifolds and one would obtain similar results, as long as the rigidity result of Ewing-Liulevicius still holds (cf. [8]). In particular, one can formulate and prove our Main Theorem for actions on the standard flag manifold $U(n)/U(1)^n$.

REFERENCES

[1] M. Artin and B. Mazur: Etale homotopy. Lecture Notes in Math. 100, Springer, 1969.

[2] M. F. Atiyah and D. O. Tall: Group representations, λ-rings and the J-homomorphism. Topology 8 (1969), 253-297.

[3] A. K. Bousfield and D. Kan: Homotopy limits, completions and localizations. Lecture Notes in Math. 304, Springer, 1972.

[4] S. Brewster: Automorphisms of the cohomology ring of finite Grassmann manifolds. Ph. D. Dissertation, The Ohio State University, 1978.

[5] S. Brewster and W. Homer: Rational automorphisms of Grassmann manifolds (to appear in PAMS).

[6] L. Dornhoff: Group representation theory. Marcel Dekker Inc., 1971.

[7] B. Eckmann and G. Mislin: Chern classes of group representations over a number field. Compositio Mathematica 44 (1981), 41-65.

[8] J. Ewing and A. Liulevicius: Homotopy rigidity of linear actions on friendly homogeneous spaces. J. of Pure and Appl. Algebra 18 (1980), 259-267.

[9] E. Friedlander: Maps between localized homogeneous spaces. Topology 16 (1977), 205-216.

[10] H. Glover and W. Homer: Endomorphisms of the cohomology ring of finite Grassmann manifolds. Lecture Notes in Math. 657, Springer 1978, 170-193.

[11] H. Glover and G. Mislin: On the genus of generalized flag manifolds. L'Enseignement mathématique XXVII (1981), 211-219.

[12] M. Hoffman: Cohomology endomorphisms of complex flag manifolds. Ph. D. Dissertation, MIT, 1981.

[13] S. Illman: Equivariant singular homology and cohomology for actions of compact Lie groups. Lecture Notes in Math. 298, Springer 1972, 403-415.

[14] C. N. Lee and A. G. Wasserman: On the groups JO(G). Memoirs AMS, Vol. 159, 1975.

[15] A. Liulevicius: Homotopy rigidity of linear actions: Characters tell all. Bull. AMS 84 (1978), 213-221.

[16] A. Liulevicius: Equivariant K-theory and homotopy rigidity. Lecture Notes in Math. 788, Springer 1979, 340-358.

[17] J. Slominska: On the equivariant Chern homomorphism. Bulletin de l'Académie Polonaise des Sciences (Ser. math., astr. et phys.) 24 (1976), 909-913.

[18] V. P. Snaith: On the Künneth formula spectral sequence in equivariant K-theory. Proc. Camb. Phil. Soc. 72 (1972), 167-177.

Printed in the United States
By Bookmasters